CHILDREN THE INVISIBLE SOLDIERS

by Rachel Brett and Margaret McCallin

(Of the Quaker United Nations Office, Geneva, and
the International Catholic Child Bureau
on behalf of the Child Soldiers Research Project)

RÄDDA BARNEN
SWEDISH SAVE THE CHILDREN

Rädda Barnen (Swedish Save the Children) is a non-governmental organisation that fights for the rights of children, in Sweden and around the world. Activities are designed to improve conditions for children at risk. Rädda Barnen acts by itself and in co-operation with others by:

• identifying and analysing problems and potential courses of action;

• sponsoring practical development and support programmes, and sharing the experience gained;

• influencing public opinion.

Rädda Barnen Publishing's books are primarily for people who work with children. Our goal is to disseminate knowledge concerning the situation of children and to provide guidance and impetus for new ideas.

ISBN 91-88726-56-8

Production management: Carl von Essen, AnnaLena Andrews
Editor: Anna Dahlin
Editorial assistant: Lisbeth Segerlund
Graphic design: Stefan Lundström, Neo Media
Cover: Stefan Lundström, Neo Media
Cover photo: Gamma/IBL
First edition: 1
Printed by Grafiska Punkten, Växjö 1996

Contents

Preface

As shown in this book, the child soldiers phenomenon appears in all corners of the world. South Africa has not been spared either. It has been awful to watch children dancing round a "necklaced" corpse.

It is immoral that adults should want children to fight their wars for them. Children should be playing, not being pawns in dangerous adult games.

From a Christian perspective, children bearing arms is particularly challenging, for the Biblical concept of the Kingdom is of universal peace, when they will beat their swords into ploughshares and their spears into pruning hooks and they shall learn war no more.

We must not close our eyes to the fact that child soldiers are both victims and perpetrators. They sometimes carry out the most barbaric acts of violence. But no matter what the child is guilty of, the main responsibility lies with us, the adults. Children are easily coerced into doing things they would never have done in a normal situation.

To stop this abuse, no child under 18 should be recruited for or participate in war – whatever the wishes of the child. There is simply no excuse, no acceptable argument for arming children.

The present book deepens our knowledge about an urgent issue, and points towards solutions. It is my profound hope that it will be read, in particular, by those whose decisions can make a difference for the individual child.

The Most Reverend
DESMOND M. TUTU
Archbishop Emeritus

Foreword

OUR IMAGE OF childhood – a time of playing, learning, making friends – contrasts sharply with the experience of many of this book's children: brutalised, betrayed, learning to kill.

Recruiting children for war goes against the core concepts of the UN Convention on the Rights of the Child. It is certainly against the best interests of the child – Article 6 of the Convention – and is just as fundamentally wrong as torture.

Rädda Barnen – Swedish Save the Children – has been involved in the child soldier issue for several years, both at field level and through advocacy. We and our partners in the Save the Children Alliance have carried out rehabilitation programmes for former child soldiers, and several of our field offices in conflict areas have been confronted with child recruitment.

In the 1980s, when the Convention on the Rights of the Child was negotiated, we fought to achieve a higher minimum age for recruitment and participation than the current 15 years, and we are campaigning for an optional protocol to the Convention stipulating 18 years. To this end, Rädda Barnen publishes a quarterly newsletter on child soldiers, "Children of War", and operates a database with information on the issue.

We have noticed that although the problem of child soldiers is known among people interested in development issues and international affairs, not very much is done, and sometimes child recruitment is even condoned. As the title of the book suggests, child soldiers are often forgotten.

But this book shows that the recruitment of children is far more pervasive than is generally realised. Mostly, children are not deliberately sought as soldiers, but soldiers are needed and children are easier to recruit than adults. Furthermore, the development and proliferation of lightweight automatic weapons has made it possible for children to serve as combatants on almost an equal footing with adults.

In many situations, children are press-ganged by both governments and armed opposition groups. Many children also volunteer in search of physical or economic security, or for cultural or ideological reasons. But the single major cause of children volunteering armed opposition groups is their own experience of ill-treatment, including torture, by government armed forces, deaths of family members and destruction of homes.

Who are they, these child soldiers? Overwhelmingly, the children are from the poorest and most marginalised sectors of society. Most vulnerable of all are those without families or with disrupted family backgrounds, with no access to education or a reasonable standard of living, and those from the conflict zones themselves.

Once child recruitment starts, it tends to escalate, leading to ever more children becoming involved at an increasingly younger age. And once recruited, the children are brutalised, indoctrinated into violence, supplied with drugs and alcohol, and used for hazardous activities such as clearing mines and spying.

At every stage of the process, they are liable to injury and death: if they try to avoid recruitment or to desert, if they cannot keep up or fail to obey orders, if they are injured... And at the end of the conflict, what then? They are betrayed again – as yet no peace agreement has recognised the role and needs of child soldiers.

Although the picture portrayed in this book is disturbing and depressing, it is not without hope. Not only is there some informed military opinion against the use of child soldiers, but many families also go to considerable lengths to try to protect their children. The book's recommendations also demonstrate that there are steps which every individual and group in society could take to help reduce the use of children in warfare, and to address the consequences of such participation for the children and their societies.

As such, this book is designed for all those who care about children, in particular for individuals and organisations who are present where children are or may be recruited.

The book is the result of the collaboration of a number of organisations and the UN Study on the Impact of Armed Conflict on Children (the Machel Study). We would especially like to thank the organisations that carried out the 26 country case studies, and the authors, Rachel Brett and

Margaret McCallin, for their tireless and dedicated work of analysing the case studies and writing a highly readable and challenging book.

We hope that the book will help to stop the use of child soldiers. All concerned individuals and organisations are invited to participate in the effort.

Lennart Lindgren
SECRETARY GENERAL
Rädda Barnen

Carl von Essen
PROGRAMME OFFICER
Rädda Barnen

Introduction

THE IDEA FOR THIS project was born in the NGO Sub-Group on Refugee Children and Children in Armed Conflict (one of the Sub-Groups of the NGO Group for the Convention on the Rights of the Child, in Geneva). A Steering Group was formed with the following members: the Quaker United Nations Office in Geneva, the International Catholic Child Bureau, Rädda Barnen (Swedish Save the Children), the Henry Dunant Institute, the Lutheran World Federation, World Vision International and the research component of the UN Study on the Impact of Armed Conflict on Children (the Machel Study). The Quaker UN Office led the project assisted by the International Catholic Child Bureau.

Funding was provided by Rädda Barnen, the Machel Study, the Swiss Federal Department of Foreign Affairs, the World Council of Churches and the Lutheran World Federation. Major contributions in kind were made by the Quaker UN Office and its parent bodies (Quaker Peace & Service and Friends World Committee for Consultation), the International Catholic Child Bureau, World Vision International and the National Inter-Religious Service Board for Conscientious Objectors (USA).

The authors wish to thank all those who carried out the research, and offered advice, encouragement and assistance in many ways, in what was truly a co-operative venture involving many different partners. Particular thanks for major contributions go to Derek Brett (for the tables in Chapter 1), Rhona O'Shea (who worked as a volunteer on the Child Soldiers Research Project) and to Penny McMillin of the Quaker UN Office. In this context, tribute must also be paid to the decision of the Quaker Meeting in Geneva, endorsed by the 1979 Triennial Meeting of the Friends World Committee for Consultation, to take up the issue of child soldiers and to Dorothea Woods (of the Geneva Meeting) and to Martin Macpherson (of the Quaker UN Office in Geneva) who played a remarkable part in putting and keeping the issue on the international agenda.

The research comprised case studies of particular situations where children are or have been active participants in armed conflicts or have been recruited into armed forces. The case studies were undertaken by local non-governmental organisations (NGOs), international NGOs working in the field, government or other agencies, on the basis of questionnaires prepared by the Steering Group. Due to the attitude of the authorities or to a shortage of time and funding, it was not possible to cover all those situations in which child soldiers are known to have been involved: secondary sources have been used to fill some of the gaps.

Case studies were received for Afghanistan, Bhutan, Burma/Myanmar, Burundi, Cambodia, Colombia, El Salvador, Ethiopia, Guatemala, Honduras, Lebanon, Liberia, Mozambique, Nicaragua, the Occupied Territories (Intifada), Paraguay, Peru, the Philippines, the Russian Federation (Chechnya)[1], Rwanda, South Africa, Sri Lanka, Turkey, Uganda, the United Kingdom (Northern Ireland) and the former Yugoslavia. In addition, the governments of Australia, the Netherlands, New Zealand and the United Kingdom provided information about their recruitment and deployment of under-18s.

Eleven of the case studies were in-depth ones, responding to both questionnaires A and B; the others either addressed questionnaire A only or were negotiated individually to address the particular situation in the country. The questionnaires are reproduced in Annex 1. The case studies themselves are available from the Rädda Barnen Project on Child Soldiers (see address in Annex 4).

The purpose of this book is to develop a better understanding of the causes and consequences of children's participation in armed conflicts, not to expose or stigmatise the policies or practices of particular governments

1 The question of child soldiers was included in H. Balian, "Armed Conflict in Chechnya, its Impact on Children" (Covcas Center for Law and Conflict Resolution, Arlington, VA, November 1995), report prepared for the UN Study on the Impact of Armed Conflict on Children.

or armed opposition groups. Therefore, the material is treated in a way which is intended to avoid, as far as possible, questions about the accuracy of the presentation of particular situations, incidents or details. Instead, the focus is on what emerges as a remarkably similar pattern across different countries, regions and cultures.

To some extent, this generalisation makes a virtue of necessity. The contract to undertake this project was signed in September 1995, with a deadline of April 1996 for submission of the report to the UN Study. During this period, the questionnaires had to be developed, the countries and case writers identified, the cases researched and written up, and the report written. All participants were aware of the limitations imposed by this timetable, especially the case writers who demonstrated great understanding and flexibility in adapting their original plans to the exigencies of the time available. It is clear, however, that no individual case study is a comprehensive analysis of the situation of child soldiers in a particular country or conflict nor, taken together, do they constitute a comprehensive coverage of the situation of child soldiers in the world.

The considerations outlined above have influenced the way in which the case study material has been handled in the text. Where the text of the case studies is quoted directly, it is duly acknowledged but, in general, wording which identifies the specific situation or names of individuals, groups or places has been edited out. No attempt at concealment is intended by this practice; the missing names will often be evident to those familiar with the country concerned. The aim is to avoid the general picture being obscured by the specific instances used to illustrate it.

Another feature of the style of treating the material arises from the same considerations. No distinction is made between time periods nor between actors unless specific contrasts are being made. None of the case studies deals with remote history; all of the experiences reported are from the recent past. In some instances practices are known, or believed, to have changed; a number of the conflicts have allegedly or actually ended; several have entered a new phase between the commissioning of the case studies and the completion of this book, but the nature of the experiences reported remains contemporary, therefore they are reported in the present tense.

DEFINITIONS

At certain points there are clear distinctions to be drawn between what for the purposes of this book are defined as "government armed forces" and "armed opposition groups". The former applies to all regular governmental, irregular governmental, government-supported or condoned forces, armed police or internal security forces, militias and self-defence groups[2] although, where possible and relevant, distinctions are made between them.

Similarly the term "armed opposition group" is used to refer to all other armed groups, even where more than one armed group was operating concurrently or consecutively, and even if the group in question subsequently became the government. The distinction between government and opposition relates to legal authority and the question of territorial control at the time in question. The authors of this book take no position about the merits of the conflicts or their causes, being concerned only with the nature and effects of the involvement of children, by any party and for any reason.

Graça Machel, the expert appointed to undertake the UN Study on the Impact of Armed Conflict on Children, has stated that 18 years should be the minimum age for recruitment and participation in hostilities, in line with the general age of majority contained in Article 1 of the 1989 UN Convention on the Rights of the Child. For the purpose of this project, this was, therefore, taken as axiomatic. Thus the term "child" is used for any person under the age of 18 years although, under Article 38 of the Convention on the Rights of the Child, the minimum age for recruitment into armed forces, as well as for participation in hostilities, is 15 years. The same age is stated in the 1977 Additional Protocols to the 1949 Geneva Conventions governing the conduct of armed conflicts. Under national legislation, the minimum ages for compulsory and/or voluntary recruitment vary: the phrase "under age" is used in relation to this recruitment age, or the age formally stated in the policy of the armed opposition group con-

[2] Except where there is clear evidence that these are neither government supported nor condoned.

cerned. It should also be noted that account is taken of the situation of those now over 18 years but who fell within this age group at the time of their recruitment.

The term "soldier" is used for a member of any kind of regular or irregular armed force or armed group in any capacity and those accompanying such groups, even if not defined or identified as members, other than purely as family members; it therefore includes cooks, porters, messengers, and so on. However, a distinction is made between those who carry arms and participate in combat directly ("combatants"), and those serving in support capacities.

CONTEXT

Although some peacetime recruitment of children is included, most of the situations studied in this book are non-international conflicts. In most of these, the government denies that it is involved in an internal armed conflict as classified by the 1949 Geneva Conventions or the 1977 Additional Protocols, and therefore denies the applicability of international humanitarian law. The consequences of this denial include lack of access by the International Committee of the Red Cross (although some governments permit the ICRC to visit detainees while denying that there is an armed conflict), and that the government classifies opponents as criminals and terrorists (while in many cases freely killing members of armed opposition groups contrary to the right to life under international human rights law).

In particular, the effect on children is that they may also be treated as criminals or terrorists, subject to the same emergency legislation as adults with no account taken of their age in terms of treatment or punishment. Furthermore, the non-application of international humanitarian law means that armed opposition groups are not bound by any rules. This does not, however, absolve the government from its obligations under national law, any international human rights treaties to which it is a party (in particular in this context the Convention on the Rights of the Child) and customary international law. The apparent evidence of better treatment of recruits by armed opposition groups which invoke Additional Protocol II to the Geneva

Conventions underlines the importance of governments becoming parties to the Protocol and governments and armed opposition groups being encouraged to accept its provisions.[3]

States parties to the Geneva Conventions and Protocols, and to the Convention on the Rights of the Child, are required to "respect and to ensure respect" for international humanitarian law.[4] Knowingly to recruit, or to allow other states or armed opposition groups to recruit, those who are likely to commit atrocities raises questions about the compliance of states with this obligation. Since there is evidence that the greater suggestibility of children and the degree to which they can be normalised into violence means that child soldiers are more likely to commit atrocities than adults, all states parties to the above treaties should consider their responsibility to prevent the recruitment and use of children in this way.

Governments, armed opposition groups, the communities themselves and non-governmental organisations (NGOs) often deny the involvement of children even when the presence of child soldiers has been documented, or claim that these were isolated cases. In some instances, this made it impossible to undertake case studies. The fact of denial implies that, at least retrospectively, doubt is felt about the acceptance or encouragement of child participation. Unfortunately, its consequence is that information is more difficult to come by, thus making a greater understanding of the causes and consequences of such participation harder, as well as leaving individuals, their communities and their societies without help in addressing the experiences through which they have gone. Denial at the time has the further consequence that the situation of child soldiers is not considered or subject to debate or monitoring because their participation is unacknowledged, undocumented or falsified.

Many of these conflicts have their roots in poverty, economic and social

[3] See also F. J. Hampson, "Legal Protection Afforded to Children under International Humanitarian Law" (University of Essex, UK, May 1996), report prepared for the UN Study on the Impact of Armed Conflict on Children

[4] Article 1(1) of the four 1949 Geneva Conventions and Additional Protocol I, and Article 38(1) of the Convention on the Rights of the Child

injustice. At the same time, it is the poor themselves who pay the highest price – often conscripted or forcibly recruited into government armed forces, civil defence forces, and armed opposition groups. They are unable to pay off recruiters, or are themselves paid by those better off to send their own children as substitutes.

Those who join voluntarily often do so as one of the few ways to earn a living or support their families, or simply to survive. Without the possibility of education or employment, where else can they turn? The conflicts take place predominantly in poor rural areas, thus impoverishing them even more, making education and employment even less likely, destroying and displacing families and leaving the children even more vulnerable both to volunteering and to forcible recruitment. In the circumstances, is it surprising that children are often lured and intoxicated by the power and security of a uniform and a gun, and that this often leads them into a life of violent crime? This may be during the conflict or afterwards: reports include burglary, banditry, kidnapping, ambushing passers-by, armed robbery, theft, looting, rape, narcotic abuse and illegal gambling. Others, having learned to kill, graduate at a young age to become contract killers hiring themselves out to the highest bidder.

Moreover, the problems are not just of the individual child soldier or ex-soldier concerned. The greater the number of present and former child soldiers, and the younger they were at first recruitment, the more serious the implications for the whole society, recognising that many of them will leave the armed forces or armed opposition groups as adults. A large part of a generation may have missed out on education, on the opportunity to acquire social and economically-useful skills. There is also the question of what they have gained in the course of their "service": many of them have suffered treatment that no civilised person could countenance and have learned the value of a gun as a means of livelihood and to get what they want. No society can afford not to take steps to address such a potential threat to its peace, security, stability and the rule of law.

TABLE 1
The Case Studies

Country	*Years covered*	*Forces and Groups Studied*
Afghanistan	1978-	(1979–92) Government forces; (1979–92) Mujahideen; (1993 –) various factions
Bhutan	1990-	Government Army (RBA), Police (RBP), Militia
Burma/Myanmar	*1988-	Government forces; Opposition groups: KNU/KNLA; ABSDF; NMSP/MNLA; USWP/USWA; PDF
Burundi	1993-	Government forces; Hutu and Tutsi Opposition groups
Cambodia	1970-	successive Government forces; Opposition Khmer Rouge
Colombia	1965 -	Government forces (army, paramilitary); Opposition groups (FARC, ELN, EPL, Quintin Lame, M19); "Popular Militias"
El Salvador	1980–1992	Government forces (FAES); Opposition (FMLN)
Ethiopia	1974–1992	(to 1991) Government forces (Derg); Ethiopian People's Revolutionary Democratic Front (EPRDF); (1992) Oromo Liberation Front (OLF)
Guatemala	1960-	Government forces (Army, PAC Militias); Guatemalan National Revolutionary Unity (URNG)
Honduras	*1977-	Government forces; unnamed Opposition groups
Lebanon	1975-	Government forces; sectional militias; South Lebanon Army; PLO; Hezbollah; Black September Brigades *etc.*
Liberia	1987–1993	Government forces; Opposition NPFL, ULIMO and other factions
Mozambique	1976–1992	Government (Frelimo) Army and allied self-defence teams/local armed groups, e.g. NAPARAMAS; Opposition RENAMO
Nicaragua	1977–1979	Government forces (National Guard)
Occupied Territories (Intifada)	1987–1993	Unstructured opposition

Paraguay	*1989-	Government forces (Army, Police)
Peru	1980-	Government forces (Army, Self -Defence Committees); Opposition (Shining Path, MRTA)
Philippines	*1976-	(Opposition) New Peoples Army; also (briefly) Moro National Liberation Front (MNLF)
Russian Federation (Chechnya)	1994-	Opposition groups
Rwanda	1994-	Government forces; Interhamwe Militias
South Africa	1961-	(to 1991) Government forces; armed wings of the African National Congress and Pan African Congress (post 1991) "Self-Defence Units" & "Self-Protection Units"
Sri Lanka	1983-	Liberation Tigers of Tamil Eelam (LTTE)
Turkey	1984-	AGRK – military wing of the (Opposition) PKK
Uganda	1989-	Lords Resistance Army (LRA)
UK (Northern Ireland)	1969-	Republican and Loyalist Opposition groups; unstructured opposition
the former Yugoslavia	1991-	"Army of Republika Srpska"; "Army of Republika Srpska Krajina"; informal Serbian groups

* In these cases, the date given is of the earliest first-hand testimony quoted in the case study, the recruitment of children having gone on for an unspecified length of time previously.

Notes:

The case studies are available from Rädda Barnen's Project on Child Soldiers (see address in Annex 4).

A young Palestinian in training. The development of light arms has made it easier for children to serve as combatants.
Photo: Carol Spencer, Gamma/IBL.

CHAPTER I

The Global Picture

A sizeable proportion [of the opposition forces] were below 18 – a fact that was much noticed when the [opposition] captured [the capital] ... The subsequent demobilization ... has significantly reduced the visibility of children in the forces as those who were among the first to be demobilized were those below 18, and women.[5]

There is no complete and reliable documentation about the number of killed, wounded, disappeared and displaced persons in four years of armed conflicts ... Therefore there is no such documentation about child soldiers either.[6]

The persons interviewed claimed that lots of minors did not show up to be listed by [the UN peacekeeping mission] due to lack of confidence and fear they felt in disarming "in front of the enemy".[7]

In the world at this moment there may be a quarter of a million children under the age of eighteen serving in government armed forces or armed opposition groups. It is impossible to be more precise because most of them are "the invisible soldiers" of our title.

They are invisible because those who employ them deny their existence. No record is kept of their numbers and ages, or the ages are falsified. Many are not part of the formal strength of the forces or groups to which they are attached, but are unacknowledged servants or hangers-on. Many are enrolled in a variety of militias and citizens defence groups, the very

5 Case study for Ethiopia
6 Case study for the former Yugoslavia
7 Case study for El Salvador

existence of which is an underdocumented area, rarely included in official figures for armed forces. They are invisible because most spend their time in remote conflict zones away from public view and the prying lenses of the media. They are invisible because they tend to vanish. Many do not return from the battlefield because they are killed or, having been injured, are abandoned. When the conflict is over, they are rarely as visible among the demobilised troops as they were among the combatants at the height of hostilities. Additionally, those in their teens are invisible because they are less obviously children – requiring a second look to realise that "they are not old enough to shave". And individually they all grow older. Even if the soldier survives, in a few years the child disappears: and the child soldier becomes invisible, locked inside an adult soldier or an adult former soldier.

The participation of children in armed conflict is not a new phenomenon. In the Middle Ages a boy who hoped to become a knight would, from the age of about 14, serve as a squire, cleaning his master's armour and standing guard while his master slept. In the eighteenth and nineteenth centuries, naval warfare relied upon "powder monkeys", as young as 8 years old, to load the cannons; and on land, infantry regiments had their drummer-boys. At all times children were probably employed as lookouts, spies and messengers.

There is a continuity from these historical examples to the experience of children in armed conflict today. They are still used wherever their small size and agility are seen as assets – for espionage, communications and demining – and from many countries come reports of recruits or younger children being used as the domestic servants and household guards of officers and their families.[8]

However, modern technology has made both a qualitative and a quantitative difference to the participation of children in armed conflict. The mediaeval squire could not hope to don his master's armour until he had reached physical maturity. Even a generation ago, battlefield weapons were heavy and cumbersome. This limited children's participation to support and ancillary roles; few could serve as front-line troops. The development of modern lightweight automatic weapons – archetypically, but by no means

[8] Case studies for Burma/Myanmar, El Salvador, Liberia, Paraguay and Rwanda

exclusively, the Soviet AK47 – has transformed the capacity of children to serve as combatants on something approaching an equal footing with adults. There is an urgent need to address, internationally, regionally and nationally, the ready availability of such weapons – not only for the sake of the children but for that of the societies being torn apart by armed conflict, internal strife and violent crime.

At the same time changes in society and in expectations have also been taking place. The transition from childhood to adulthood used to take place in distinct stages. By their teens, certainly, children in most societies were expected to be making an economic contribution, but the age of full maturity and citizenship was set far higher. The latter half of this century has seen a convergence of these ages. The trend has been for the age for the franchise to come down to 18, while the duration of compulsory education has risen, so that the gap between entering the workforce and civil maturity has become considerably shorter and for many individuals has disappeared altogether. It is particularly relevant to the current study that almost all states which impose some form of compulsory military service have chosen 18 as the official age of liability (see Table 2). It is against this background that the Convention on the Rights of the Child can assume that generally until the age of 18 one is a child and thereafter an adult. The case for the objective validity of this threshold in the context of armed conflict will be made in Chapter 7.

NUMBERS AND DISTRIBUTION

> *Estimating of the number of children in the armed forces is a difficult task. We have looked at figures in casualty wards in hospitals; we have gone through a valuable resource of mine victims; we have looked at surveys on demobilisation; we have talked with staff of a major military training camp and we have discussed the numbers issue with anyone vaguely knowledgeable about the military and with all interviewees and the soldiers themselves. Not surprisingly, there is a wide variation in the estimates of the number of child soldiers involved.*[9]

9 Case study for Cambodia

Child Soldiers

In italics= participation of children under 18

In bold letters= participation of children under 15

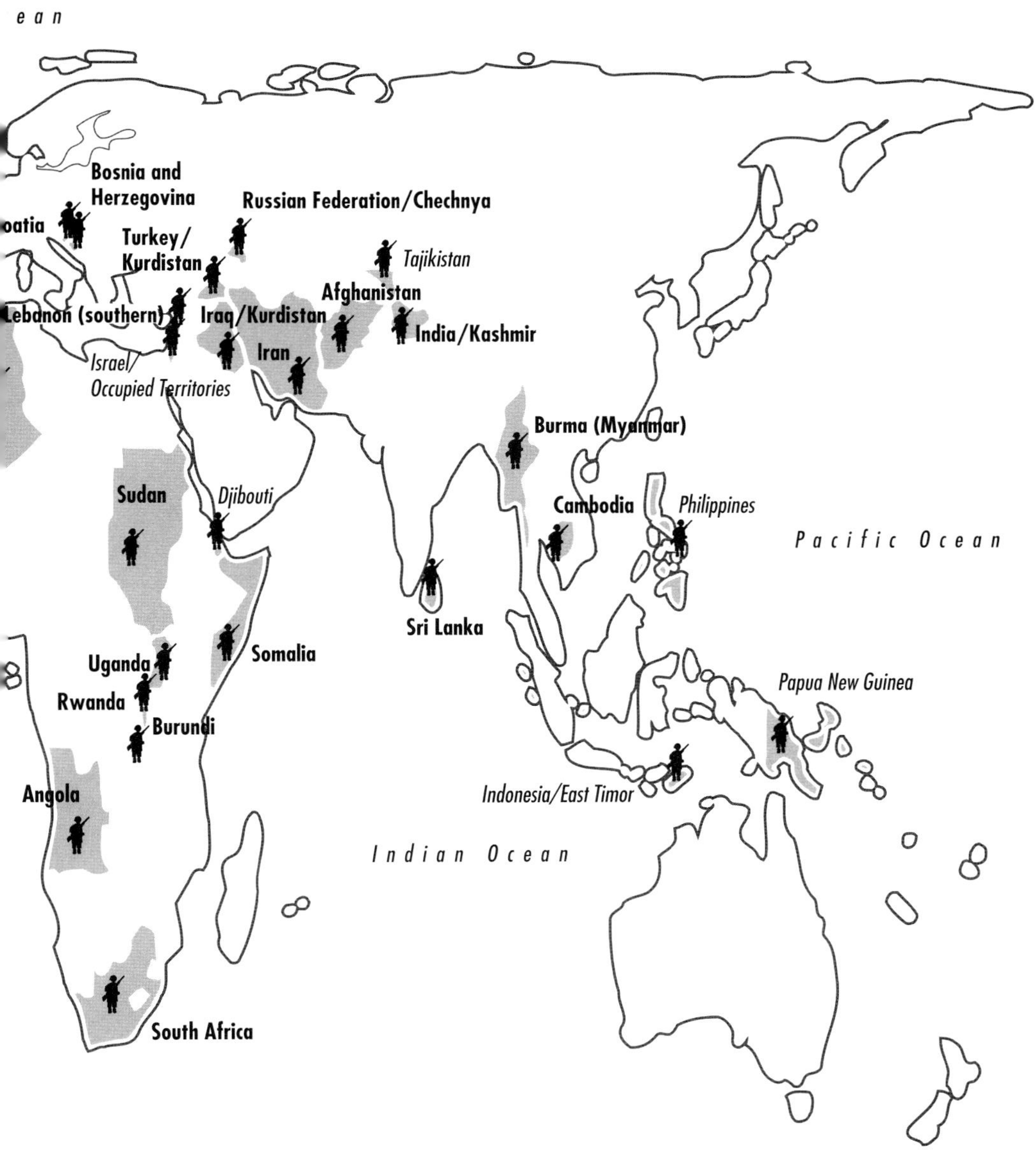

Children under 18 years of age have been reported to participate in either government or opposition forces, or both, in armed conflicts in these locations (ongoing conflicts and conflicts in which hostilities ceased during 1995 or 1996).

This information has been compiled by Rädda Barnen, Stockholm, in cooperation with the Quaker UN Office, Geneva, and Dorothea E. Woods, Geneva.

CHILDREN IN 33 CURRENT OR RECENT ARMED CONFLICTS

Children under 18 years of age have been reported to participate in either government or opposition forces or both in armed conflicts in the following locations (ongoing conflicts and conflicts in which hostilities ceased during 1995 or 1996). Bold letters indicate the participation of children under 15 (26 cases). See map on pages 26–27 for a geographic overview.

Afghanistan
Algeria
Angola
Bosnia and Herzegovina
Burma (Myanmar)
Burundi
Cambodia
Colombia
Croatia
Djibouti
Ecuador/Peru
Guatemala
Indonesia/East Timor
India/Kashmir
Iran
Iraq/Kurdistan
Israel/Occupied Territories
Lebanon (southern)
Liberia
Papua New Guinea
Peru
Philippines
Russian Federation/Chechnya
Rwanda
Sierra Leone
Somalia
South Africa
Sri Lanka
Sudan
Tajikistan
Turkey/Kurdistan
Uganda
United Kingdom/Northern Ireland

This information has been compiled by Rädda Barnen, Stockholm, in co-operation with the Quaker UN Office, Geneva, and Dorothea E. Woods, Geneva.

The accompanying map shows 33 current or recent[10] armed conflicts for which reports of the participation of children under eighteen years of age have been received by the Rädda Barnen Project on Child Soldiers. Twenty of the conflicts in question have been considered in the case studies which form the principal sources for the present book. Case studies have also been undertaken for El Salvador, Mozambique and Nicaragua, in which armed conflict ended before 1994, and for Bhutan, Honduras and Paraguay, where no armed conflict has been taking place (see Table 1 for an overview of the 26 case studies).[11]

It should be noted that the case study on the Occupied Territories deals only with the Intifada – the involvement of children in unstructured and essentially unarmed acts of politically-motivated violence – and that similar situations are mentioned in the case studies on Northern Ireland and South Africa. While at times such events have been discreetly orchestrated and co-ordinated by armed opposition groups, there is no doubt that in all three places they also occurred spontaneously, in which instances the participants could not be regarded under even the broadest of definitions as "child soldiers".

Because of the limited geographical, and in some cases historical, scope of the case studies, this book is not an exhaustive survey of child soldiers in the world today. However, as such a survey has yet to appear, Table 2 (p50) presents what estimates are available of the total number of child-soldiers as compared to the population, total strength of armed forces and, where known, armed opposition groups, and to the number of deaths in war or political violence over approximately the last thirty years in each country. Even when indirect casualties, such as famine deaths in war-torn countries, are excluded, the majority of such deaths are of civilians (conversely suicides and other deaths sustained by soldiers not in action are excluded),

10 Conflicts in which hostilities ceased during 1995 or 1996.

11 Bhutan has been classified as one of 40 "serious dispute situations" falling short of open armed conflict by A. J. Jongman & A. P. Schmid, "World Conflict Map, 1994–1995", PIOOM Newsletter and Progress Report, 7, 1 (Leiden University, 1995), p 23. The case study for Honduras documents the use of children by armed opposition groups.

but the figures still give some indication of the level of violence in the countries concerned.

Where information about the existence of compulsory military service ("conscription") is available, this is indicated in Table 2, together with the age at which citizens become liable for such service and its duration in months. Similarly, when the minimum age for voluntary recruitment into the armed forces is known, this is indicated. Conscription usually applies to all males apart from certain specially-exempted groups; in a small minority of countries it applies to females as well. However, in some countries only a proportion of the eligible population is required to serve. They may be selected by some random method (such as the draft in the USA during the Vietnam war), or by imposing a numerical quota on particular groups or communities. In some countries (for instance Mexico) there is legislative provision for the relevant names to be selected by lottery. How some such systems work in practice is shown in the chapter on recruitment. The terms of compulsory service vary considerably. It is not unusual for the duration to vary between different services; the army usually being the shortest, the navy often the longest. Further, in many countries a relatively short period of full-time training (the figure given here) is followed by many years in the reserves, which may entail brief refresher courses.

Finally, the Table shows for comparison the age for the franchise where applicable and the relevant international humanitarian law treaties to which each country is a party.

The most striking feature of Table 2 is the large expanse of blank space. This does not show the part of the world where there are no child soldiers. It shows that part of the world for which at present we have no information. Not even for all of the countries covered by case studies was it possible to obtain an estimate of the total number of child soldiers; in some cases a proportion was quoted, in others the proportion has been calculated from the estimated total. Often it is possible only to indicate by question marks whether the figures are, as far as we can judge, in the hundreds, thousands, or whatever. Similar guesses have been included for those countries which are known to legally recruit under-18s. There is evidence in the case studies, that even when confronting armed opposition groups, some government forces[12] implement recruitment procedures with a minimum age of 18 and thereby limit the child soldiers in their own ranks to a few isolated and marginal exceptions. However, about half the case studies report substantial illegal under-age recruitment into government armed forces. This gives room for doubting whether all other countries abide scrupulously by the applicable national legislation.

If we add together the estimated actual numbers of child soldiers quoted in Table 2, the total comes to about 50,000. If the "question marks" are given their minimum value (for example 100 for ?00), this adds a further 75,000+. Doubling that, which still treats the estimates very cautiously, it equates with the figure of 200,000 suggested in UNICEF's State of the World's Children 1996, without including any allowance for more than a hundred of the states and territories listed.

In fact, while interesting, such speculations are ultimately pointless. The total number of child soldiers in each country, let alone the global figure, is not only unknown but unknowable. Child soldiers are commonest in the thick of conflicts; precisely where accurate, up-to-date information is hardest to obtain. As will be seen from the chapter on recruitment, the greatest number is employed without enquiries being made as to their ages. Those who employ them and who alone are in a position to hold definitive data on numbers and ages, rarely have any incentive to obtain or collate such data and usually have a strong perceived interest in obstructing or deterring the access of third parties to such information. When those in authority are aware of existing national and international standards, they are reluctant to reveal breaches. This not only makes the collection of information harder, but may distort such information as is obtained.

Even if all these obstacles could be surmounted, the facts themselves would still remain elusive: of the former Renamo child soldiers analysed in the case study on Mozambique, 16 per cent of the boys and no less than 28 per cent of the girls did not know their own ages. In any case, as individual children are always growing older, any figures regarding child soldiers will be out of date by the time they are published. Furthermore, political or

[12] For example, the Philippines, Russian Federation and Turkey

Girls and boys are trained in Beirut 1986. Photo: Pica Pressphoto.

military developments can change the position in any specific country out of all recognition almost overnight.

GENDER

... families will hardly allow their girls to marry former child soldiers...[13]

Female guerillas can be seen these days in almost all [opposition] camps and gradually they are taking on more active duties than only cooking and/or cleaning.[14]

The girl soldiers were stationed at the front in all military actions, and so bore the brunt of any casualties.[15]

Nowhere is the influence of cultural expectations seen more strongly than in relation to the gender of child soldiers. Reports tend to mention "children" and very careful reading is sometimes necessary to ascertain whether the children are of both sexes or only boys.

The main impression given by the case studies is of a contrast between government armed forces and armed opposition groups, with the former overwhelmingly if not exclusively male, while the latter tend to involve both males and females of all ages. There are exceptions in both directions: in Peru both men and women are covered by the Law on Compulsory Military Service; in Afghanistan (see quote above) the unstated implication is that all child soldiers on all sides are boys.

In fact, the contrast between government forces and armed opposition groups is not as absolute as it appears at first sight. Whereas everyone living in a guerilla camp, except perhaps the smallest of the children, is self-identified as a member of the group, the equivalent is not necessarily true of the civilian population on which government armed forces depend. Behind the combatant forces there is usually a vast hidden army of support and ancillary workers who may or may not show up in official figures. These are often women and girls, and sometimes small children:

[13] Case study for Afghanistan
[14] Case study for Turkey
[15] Case study for Cambodia

Women ... essentially provided support services such as secretarial and musical services. More and more women, however, were involved in these services as the size of the army expanded ... As the services were far removed from combat duty and were mainly carried out in urban areas an increasing number of young girls were looking at the armed forces ... as a means of winning their living in the context of widespread unemployment.[16]

One interviewed ex-military said: "Every officer used to have his assistant or servant, who was a cipote [little boy]." ... In general, women and female children ... performed "domestic functions" like cooking, washing, cleaning, doing secretarial work in administration etc.[17]

In Cambodia, large parts of the government armed forces function in a way which would normally seem more characteristic of armed opposition groups:

... families live in temporary huts constructed near the army base where the soldier is serving ... often the families will live in the camp itself with the soldiers ... and cook for them and eat with them ... Families are also to be seen in the military and provincial hospitals caring for the wounded and sick soldiers ... Women and girls have been used in support roles ... and [one] study gives an example of a soldier wife being used as a porter in a dangerous area. Women also fight on the government side, but these women are invariably married to soldiers or have family in the force and there has been no evidence of women under 18 being recruited.[18]

This said, the fact remains that women – and girls – are more to the fore in armed opposition groups, and are more likely to be combatants:

Unlike their counterparts in the [government] army, women in the [opposition] force usually assumed combat duties. In fact, some are said to be among the best fighters. Not an insignificant number of them were

[16] Case study for Ethiopia

[17] Case study for El Salvador

[18] Case study for Cambodia

posted in commanding positions within the force ... again, not an insignificant number were believed to have been recruited while they were under age.[19]

The braver girls participated fully in war including a few who were commanders.[20]

... the commander of one of the self-defence units was a girl aged 16.[21]

Female children were not treated differently either, unless they requested to be assigned to a medical or kitchen unit, in which case the request was accommodated.[22]

For example, in El Salvador, where estimated government forces of 60,000 included only one short-lived battalion of 160 armed women and girls, of 2,000 demobilised opposition troops who were aged under 18, some 700 were girls. In Ethiopia it was estimated that women and girls formed between 25 and 30 per cent of the opposition strength. In Cambodia, on the opposition side, there is a report that "a group of 300 to 500 girls under 15 were kept together and given military training". Of 183 child soldiers recaptured from the armed opposition group in Uganda, 55 were girls. Of the ten interviews with members of the NPA opposition group in the Philippines, six were former child soldiers, and two of these were women. The involvement of women is extending to societies, such as the Kurdish population in Turkey or the Tamils in Sri Lanka, where women have traditionally played a subordinate role.

Many, many women come to the guerillas wanting to fight, a stream of them, but they in particular have been held back in their development by feudal society.[23]

In Turkey there are reports that the PKK has organised a "children's bat-

[19] Case study for Ethiopia
[20] Case study for Uganda
[21] Case study for South Africa
[22] Case study for Chechnya
[23] Case study for Turkey

talion", comprising three divisions, two of boys and one of girls. This seems to reinforce the pattern discernable in the other cases quoted, namely that when girls are fully involved, they provide approximately one in three combat troops.

Although, in general, girl soldiers perform the same functions as boys, sometimes distinctions are made:

> *Soon after recruitment the girls are divided up and allocated to rebel men to be their "wives".*[24]

> *... the [armed opposition group] also recruits young girls into roles supporting the adult recruits, in preparation of food, attending to the wounded, washing clothes, without neglecting the relevant military training.*[25]

> *Girls are also enrolled into the opposition armed group and essentially take care of the kitchen work and the rendering of sexual services, except those who ... are broken into guerilla techniques.*[26]

> *... in the [opposition] camps, even the boys, already traumatised themselves, violently sexually assaulted the girls, even threatening them with death or removal of food rations if they resisted.*[27]

> *... girls took on the traditional household chores (looking after the youngest ones, cooking ...), and the older ones were 'promoted' to marriage.*[28]

It is not clear whether marriages were forced or merely encouraged. Such marriages for young girls may complicate demobilisation, when they may be torn between staying with their husbands or returning to their families.[29] Others may be teased by their fellows if they return to school.[30]

24 Case study for Uganda
25 Case study for Guatemala (translated from the Spanish)
26 Case study for Burundi (translated from the French)
27 Case study for Mozambique (translated from the Portuguese)
28 Case study for Mozambique (translated from the Portuguese)
29 Case study for Mozambique
30 Case study for Uganda

Involvement in sexual activity for girls (and indeed boys), whether dignified as marriage or not, entails for many the additional hazards of sexually transmitted diseases, HIV/AIDS, and pregnancy. The recruitment of young girls may be a deliberate attempt to provide "wives" free from HIV infection, thus "the criteria used for 'marrying' girls to rebel men seems to be signs of puberty."[31] More than one armed opposition group requires abortions if female members become pregnant.[32]

In Sri Lanka, there has been a tendency for girls to be used as suicide bombers. This is also the case in Lebanon where, otherwise, no evidence of female involvement in the conflict was given by the case writer.

AGES

In Table 2, wherever possible the age of the youngest reported child soldier is given. Although these extreme figures draw attention to the full size – and horror – of the problem, they are rarely representative. Many are based on unconfirmed anecdotal evidence. On the other hand, they sometimes represent the only specific evidence on ages cited.

Fortunately, some of the case studies have given information from which it is possible to derive figures for the age breakdown of at least a sample of child soldiers. We also sought official information from the five countries we knew to have legal provision for the peacetime recruitment of sixteen year olds (Canada did not reply). In these cases the information is taken from the replies to enquiries made of the governments concerned through their Missions to the United Nations in Geneva. These figures are unique in the degree of their authority and because they represent recruitment which is carried out in complete conformity with the current national legislation. As the age intervals used in the case studies varied from source to source, the only way of presenting the data in a format which allows ready comparison for different countries is to graph the cumulative percentages (that is each line reaches 100 per cent at 18, to show that all child soldiers are aged under 18). This is done in Figure 1.

In interpreting this graph it is important to bear in mind that it shows

31 Case study for Uganda

32 Case studies for Honduras and Peru

FIGURE I

Age Distribution of Child Soldiers in eleven countries

In this graph, the line for each sample starts at the minimum age recorded and goes up to 100% at 18 (all child soldiers are by definition aged under 18). Half the child soldiers in each sample are below the age where the line crosses 50%, and so on.

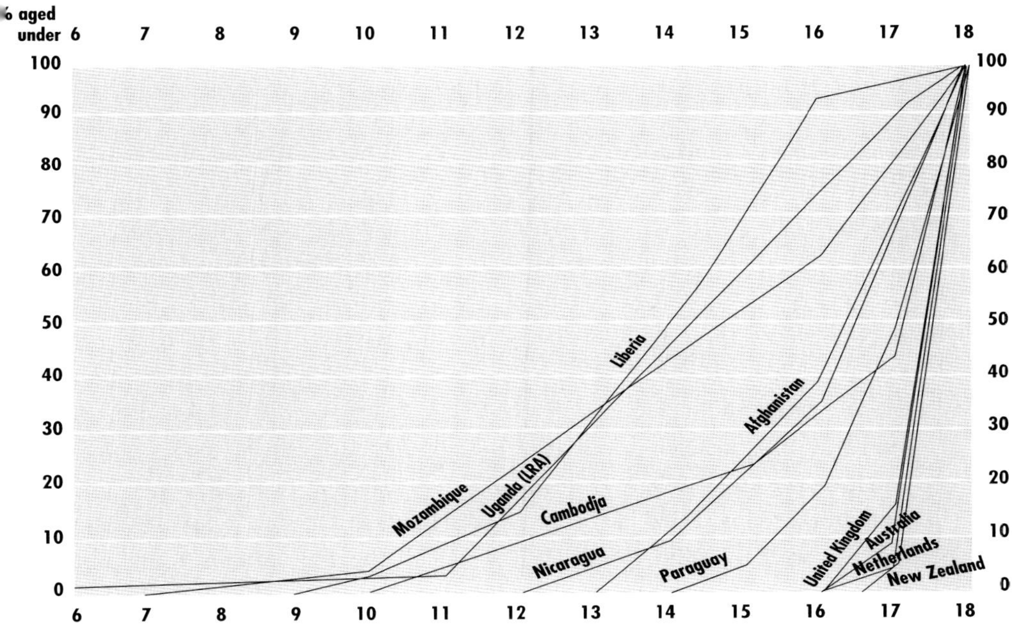

Sources:

Australia, Netherlands, New Zealand, United Kingdom: Derived from recruitment figures provided in 1996 by the governments concerned through their missions to the UN in Geneva.

All other countries: The case studies. In the case of Cambodia figures relate to the total number of child soldiers identified in three non-overlapping studies of military landmine victims 1994–96. In that of Paraguay, they are derived from a list, with ages, of the 30 soldiers alleged to have died in suspicious circumstances 1989–1995, of whom 20 were aged under 18.

the data which happen to be available and is not a representative sample of the forces and groups which employ child soldiers. Therefore, while perhaps the most immediately striking feature of the graph is that all the lines on the left-hand side (that is, those which show a high proportion of *young* child soldiers) relate to Africa, these samples are all taken from armed opposition groups, while the eight lines on the right-hand side all relate to government armed forces (although the ratios shown for Afghanistan are believed to hold good for all sides). In the three "left-hand" samples, at least half the child soldiers were aged under 15, whereas in none of the others were as many as a third of the children so young.

Even where there is strong evidence of under-age recruitment into government armed forces, the number of child soldiers increases sharply above the 15-year-old threshold set in Article 38 of the Convention on the Rights of the Child and in the 1977 Additional Protocols to the Geneva Conventions, and continues to increase up to the official national recruitment age.

The opposite is true of the sample reported in the case study for Liberia. As this related to children handed over to a demobilisation programme, however, it may not be representative of all child soldiers in the conflict. If it were, the question would arise as to what had happened to the 16 and 17 year olds; could it really be that the armed groups concerned were composed of adults (over 18) and something like 25 per cent children under 16, with hardly anyone in between? It may be that the demobilisation programme concentrated on getting the youngest children out first. It may be that the younger children were more readily identifiable. Perhaps there was a tendency to hold on to those aged over 15, whose presence could be justified in the current state of international law and/or were the demobilised soldiers those of least experience and military value?

It is notable that in an earlier study of demobilised government child soldiers in neighbouring Sierra Leone[33] a similar age distribution was recorded. However, the Sierra Leone case study gave a detailed breakdown of the length of time the children had spent as soldiers: none of the demob-

33 Case study for Sierra Leone in M. McCallin: *The reintegration of young ex-combatants into civilian life* (ILO, Geneva, 1995)

ilised children were recorded as having spent more than two years in the armed forces. Among those aged over 15 and who might have been expected to have been in the longest, the maximum service recorded was one year and nine months.

At first glance, there is no similar distortion to the age distribution recorded in the group of former child soldiers of Renamo analysed in the case study for Mozambique. The surprising feature about this group is that it is so small. The demobilising agency of the Mozambique government estimated that there were 850 former Renamo child soldiers, of whom 574 were identified and referred for rehabilitation. On even the most conservative estimates of Renamo's strength made during the civil war this would imply that no more than about four per cent were aged under 18, which contrasts oddly with most of the other figures in the case studies and the extent to which Renamo was at one time a synonym for the abusive recruitment of children.

Three possible explanations suggest themselves. The first is that the estimates made during conflict are even grosser exaggerations than one might expect. The second is that towards the end of the conflict many child soldiers may have faded away from the ranks or been informally demobilised and thus, as in El Salvador and Ethiopia, not been picked up in the demobilisation process. The third possibility is that, even while identifying former child soldiers who were no longer children, the programme may have consciously or unconsciously overlooked those who did not commence their service at a very young age. This last hypothesis tallies with those discussed in explaining the West African figures. If the 16 and 17 year olds have not been consciously excluded, however, it suggests that while the role of the very young may have been exaggerated in the past, the estimates that less than 25 per cent of these forces are made up of child soldiers overlook substantial numbers who are approaching the age of 18 and are thus less readily distinguishable from the adults.

Although the age distribution of child soldiers is the most easily available information, it is not the most useful. It is far more revealing to know, on the one hand, what proportion of the total strength of the force or group is made up of children and, on the other, the average age of recruitment. The case study for Mozambique observes that the "first" soldier demobilised by Renamo was aged 16 , but had in fact been recruited at the age of nine. Former child soldiers had spent up to 12 years with Renamo, and the ages of those referred to the rehabilitation programme ranged up to 24.

The above factors are of course linked but do not correlate directly because the third element in the equation is the average length of service. Where recruitment ages are widely ignored, but the short statutory length of service is respected, as for example in Paraguay or apparently in El Salvador, the situation will rapidly be reached where the majority of the army is aged under 18 at any one time. In the case of Paraguay, where the length of service is only one year, the case writers estimate that something like two-thirds of those performing military service were actually aged under the official recruitment age of 18.

By contrast, only two or three per cent of the overall strength of the armed forces in the United Kingdom and New Zealand are aged under 18, but the figures produced by their Ministries of Defence showed that something over 35 per cent of non-commissioned personnel in the UK ("the officer cadre are all over age 18 on commissioning") and 40–50 per cent in New Zealand are first recruited below that age. An average length of service in excess of 15 years in a voluntary, professional force accounts for the difference between the figures for Paraguay on the one hand and UK/New Zealand on the other.

Such considerations probably explain the apparent discrepancies discussed in the case study for Cambodia, between the contemporary figures for military mine casualties, only one or two per cent of which are aged under 18, and the ages at which interviewed soldiers claim to have been recruited. In one 1995 survey of 1,200 soldiers, 13.1 per cent had been recruited as children; they had however served eight years on average. Compared with the proportions elsewhere, this would seem to be rather low. More convincing, despite the smaller sample involved, is the estimate based on a survey of mine casualties in military hospitals, namely, that 43 per cent had originally been recruited at under 17 years.

El Salvador provides an example of where this feature led to a probable disparity between the proportion of child soldiers fighting in the government and the opposition ranks. Despite the fact that both parties were known to have recruited heavily among teenagers, the proportion in the opposition ranks was estimated at something above 20 per cent, while estimates of the proportion of child soldiers on the government side ranged as high as 80 per cent. It is likely that armed opposition groups, which typically require a lifetime commitment from their members, will rarely count more than 20–25 per cent of child soldiers in their ranks, whereas no such limit necessarily applies to government armed forces. Such a theory would certainly be consistent with the various estimates made in the case studies.

With regard to recruitment ages, it is interesting that among the former child soldiers in armed opposition groups interviewed for the case study on Burma/Myanmar, the most common age of recruitment was 14, followed by 17, and the youngest age was 10. By contrast, of those who had served on the government side, the most common ages were 16 and 17, only 3 out of 20 had been recruited at 14, and none younger. Although this is a very small group and not a random sample, it is tempting to read into these figures some evidence that much enlistment into the opposition force is to pre-empt the danger of conscription into the government armed forces (a feature which will be discussed in the chapter on recruitment).

CHANGES OVER TIME

> *... literally thousands of youths would have been involved in such riot situations over the last 25 years.*[34]

The first international regulation of the use of children in armed conflict was the establishment of a minimum age for recruitment in the 1977 Additional Protocols to the Geneva Conventions. But it was in January 1986 that international attention was dramatically focused on child soldiers in their modern form, when Yoweri Museveni's National Resistance Army fought its way into Kampala, the capital of Uganda, and was seen to have four- and five-year-olds in its ranks. It was not that Museveni's were the first modern child soldiers. They simply, and uncharacteristically, became visible by marching into the capital. Iran, for example, was already widely reported to be using boys as young as nine in "human wave" attacks

[34] Case study for Northern Ireland

in its war with Iraq, and in many countries the armies had for generations been filling their ranks by the forced recruitment of youngsters, as will be discussed at greater length in the following chapter.

Nevertheless, it would certainly seem that the scale of the problem has been growing in most places over recent decades. Even in those case studies which deal with countries where continuous or intermittent armed conflict has been going on for many years, there is evidence that the real upsurge is a relatively recent development.

Since the 1949 war of independence, Burma/Myanmar has faced a series of conflicts. However, the case study is based on interviews with young people of various ages, all of whom joined armed opposition groups during or since the pro-democracy uprising of 1988, and defectors who had been recruited into the government armed forces after that date. Some of the testimony quoted in the case study indicates that it was only after 1988 that the recruitment of under-18s by government forces began.

In Guatemala, armed conflict has been going on since the 1960s, but the PAC militias, in which it seems likely that the greatest number of boys are included, were not founded until the early 1980s. The various Colombian conflicts are of an equally long-standing nature, but reports from early 1996 quoted in the case study indicate that the major armed opposition group may recently have more than quadrupled its strength.

In the case study for Cambodia, it is reported that there is no evidence of the recruitment of significant numbers of children in the 1960s, but that "there is widespread evidence of vast numbers of child soldiers being used during the Lon Nol period [1970–75]", with a peak in recruitment in 1973–74. Child soldiers have featured prominently in all subsequent stages of the Cambodian conflicts.

Of the case studies, that for Nicaragua deals with the events which are most distant in time, covering a group of 92 child recruits to the National Guard selected from about 200 who were captured following the defeat of the Somoza government in the civil war of 1976–79 - a group who now will be aged in their thirties. No indication is given of the total size of the force, or of the numbers or proportion of child soldiers who fought on the other side. However, it is known that of 22,000 irregular combatants demobilised after a further ten years of conflict in Nicaragua, 1,800 were less than 16 years old.[35] The problem had certainly not declined.

In Afghanistan, the official recruitment age was 22. After the outbreak of the civil war in 1978 it was reduced to 20, to compensate for defections to the ranks of the Mujahideen, and eventually was reduced to 15. By the time the Najibullah regime fell in 1992, it is estimated that some 26 per cent of its force was aged under 18, including ten per cent aged under 16. A very slightly higher proportion of the Mujahideen forces were under 18. However, while the ratio between the under-16s and the under-18s has remained more or less constant, the case study writers estimate that no less than 45 per cent of the strength of the various factions which have emerged since 1992 is aged under 18. Furthermore, the proportion of front-line fighters among the children has increased to two-thirds.

A parallel sequence of events has taken place in Ethiopia. There, a Marxist coup in 1974 was followed by the growth of a number of political and regional armed opposition groups which mobilised a broad spectrum of the population, including children. Major recruitment drives were held by the government in 1977, 1982 and 1990. The degree of compulsion and the comparative youth of those targeted increased each time. By 1988, one report spoke of 50 boys aged under 16 in a prisoner of war camp containing 1,500 government troops. By the time the government fell in 1991, as many as 15 per cent of its troops may have been under 18, and the proportion of children in the victorious opposition forces may have been as high as a quarter. From both sides it was the youngest soldiers and the most recent recruits – usually the same thing – who were demobilised most hastily; thus the evidence vanished. As in Afghanistan, the victory of one armed opposition faction did not herald the end of the conflict, which has continued on a number of fronts. The case study does not bring the picture up to date, but it does report, that in 1992, the defeat of one faction which had split from the governing coalition resulted in the capture of 22,000 opposition troops of whom 8,490, or just under 39 per cent, were aged under 18.

35 Source: UNICEF 1992, quoted in J. M. Tortorici, "Peace Education in Nicaragua" in M. McCallin (Ed.): *The Psychological Well-Being of Refugee Children. Research, Practice and Policy Issues* (ICCB, Geneva, 2nd edition 1996)

In Liberia, too, the apparent victory of one opposition faction followed by a peace accord and the beginnings of a demobilisation programme has, in the event, seemed simply to herald a further splintering of factions and a new and even more destructive phase of the conflict. The case study in fact relates to an early stage in this process, but it is notable that the estimates over time of the proportion of children involved in the various groups, and of the extent to which they play an active role, have risen.

It is essential not to fall into the trap of thinking of a child soldier as a static phenomenon. In fact, the individual child soldier is essentially transitory. Within a few years he or she will be an adult soldier, a former soldier, or dead. At times this may seem irrelevant; although the individuals change, the process continues. However, as the recruitment of child soldiers progresses, so its effect on the societies concerned is cumulative. There is rarely enough information on which to construct even a rough estimate of this cumulative effect. However, the civil war in El Salvador was a long-drawn-out and indecisive struggle. From about 1982 to 1990, it was reported that the government armed forces were recruiting between 12,000 and 20,000 youths annually, almost all below the official age of 18. The recruitment figures would imply that in a country of five and a half million, there have been well over 100,000 who participated in the war as child soldiers on the government side alone. How many have survived? Where are they now? Indeed, how many of the age group affected avoided the fate of becoming a child soldier on one side or the other? (*Who* avoids service is addressed in the chapter on recruitment.)

Instances where the number of child soldiers has declined are rare. The effective end of the conflicts in El Salvador and Mozambique are two obvious examples. In Peru the, at least temporary, defeat of the Shining Path has presumably led to some decrease in the number of child soldiers. Of conflicts not covered by case studies the Iran/Iraq war was best known for the use of child soldiers and its end considerably reduced the scale of the problem.

The one dramatic positive change recorded in the case studies has come with the change of government in South Africa, but that is associated with a disturbing development. The positive change is that conscription, which used to take place at the age of 16 (for white males) has been abolished in favour of an all-volunteer force with a minimum age of 17. Despite the amalgamation of former government and opposition forces, this has through attrition and normal aging led to a rapid reduction in the number of child soldiers. At the same time, the case study documents the large number of young people who have organised themselves into armed "Self-Defence Units" and "Self-Protection Units" which are not considered as opposition groups but are not effectively answerable to any civil authority.

A parallel development, once again involving groups of almost exclusively children, which claim to be community based, but whose allegiance is uncertain, is reported from Colombia: "popular militias" – "groups of youths between 14 and 17 years old who engage in self-defence of the population ... [which] consider themselves the armed wing of the community".[36]

An increasing number of factions with an ever-younger armed membership and the development of armed groups of children whose claim to legitimacy is unclear suggest a downward spiral. The analysis in the case study for Afghanistan points to a strong association between the increasing number and decreasing age of the children involved in the conflict and the destruction of the economy, but most strongly with the almost complete loss of educational facilities. The case writers estimate that some 90 per cent of the country's children now have no access to schooling. Even more significant are the wider implications of their observation:

> *The continuous involvement of child soldiers in war can be a cause of further deterioration of the security situation in future. This may even result in a situation that, in reality, peace will not show its face for decades.*[37]

This is by no means the only one of the situations covered by the case studies where a similar question could be posed about the prospects for peace in the light of the fact that children are growing up without education, trained in combat and their lives' meaning is derived from their participation in the conflict. Thus the extensive involvement of children as combatants may in itself be a significant factor in prolonging the conflict.

[36] However, the case writer comments, "at present some sectors of the militias may have relations with guerilla groups".

[37] Case study for Afghanistan

Young recruit in the army of Paraguay. Several countries still admit recruits from the age of 16, including Canada, Colombia, New Zealand, Paraguay and the UK.
Photo: Jorge Sáenz.

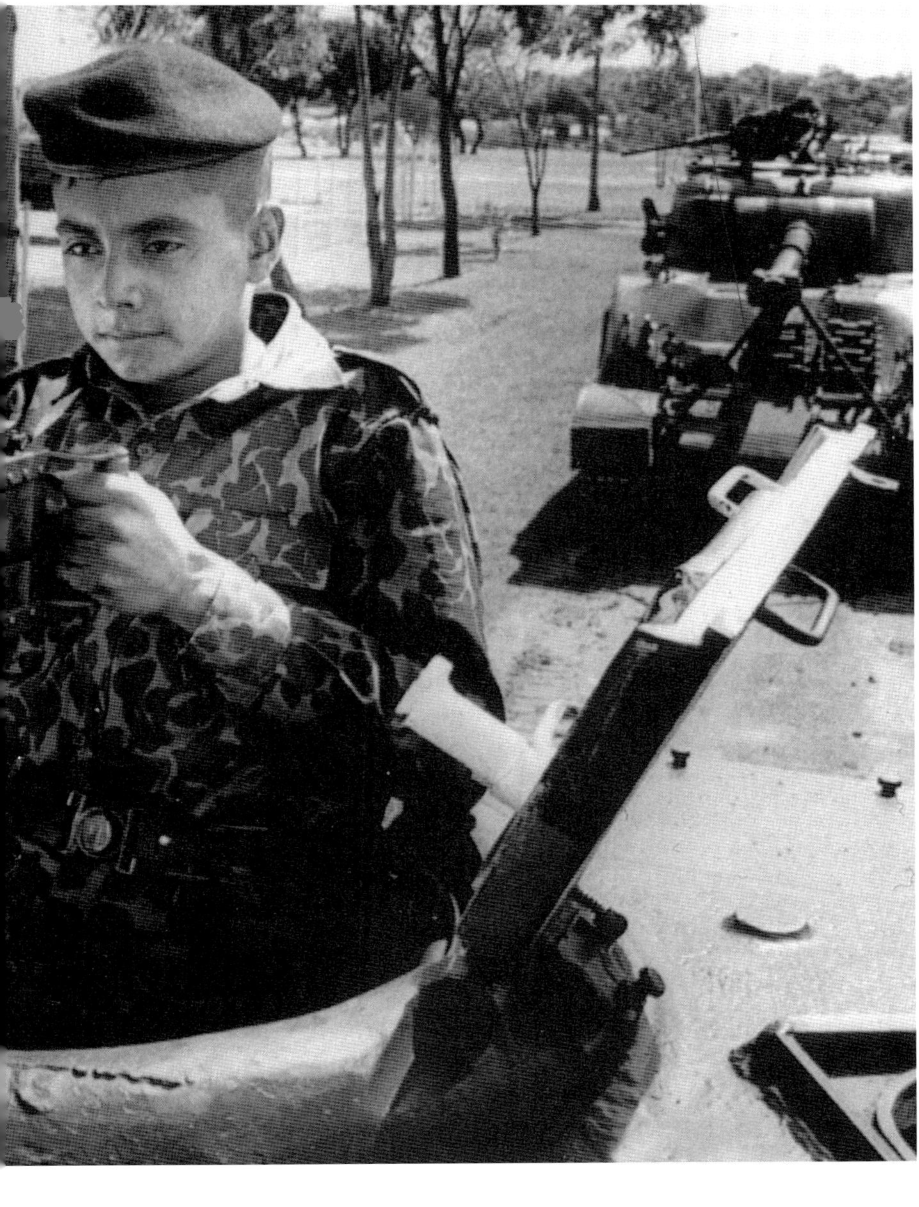

TABLE 2

Child Soldiers in the World

GENERAL NOTES

Footnotes expanding or explaining the individual country entries are printed at the end of the Table on page 62 et seq.

Sources (with the shorthand used to refer to them in footnotes)

"Case Study": Individual country case study prepared for the present book. (The source, unless another is specified, for data on child soldiers.)

"Goodwin-Gill & Cohn": Goodwin-Gill, G. & Cohn, I., *Child Soldiers*, Oxford University Press, 1994. (The source, unless another is specified, for information on existence of conscription and legal recruitment ages.)

"Guinness": *Guinness World Fact Book*, Guinness Publishing, London, 1994. (The source, unless another is specified, for information on populations, government armed forces, and lengths of compulsory military service. Also the source of references to selective conscription.)

"ICRC": International Committee of the Red Cross, World-Wide Web home page on Internet. (The source for information on ratifications to the Geneva conventions and protocols, as of 31 May 1996).

"IPU": Inter-Parliamentary Union, Geneva, printed list on voting ages, dated April 23, 1996.

"SIPRI": Sollenberg, M. & Wallenstein, P., "Major Armed Conflicts", in *SIPRI Yearbook 1995*, Oxford University Press, New York, 1995, pp 26–36. (The source, unless another is specified, for information on the size of armed opposition groups.)

Among other sources consulted were: Jongman, A. J. & Schmid, A. P., "Contemporary Conflicts" and "World Conflict Map 1994–95", in *PIOOM Newsletter and Progress Report*, Vol. 7, No. 1, (Leiden University, 1996), pp 14–23, and Project Ploughshares, *Armed Conflicts Report 1995* (Institute of Peace & Conflict Studies, Waterloo, Ontario)

Explanations to symbols, abbreviations, etc:

- None, or not applicable. (In column H less than 0.1 per thousand). Blank spaces mean that no information is to hand. They do not mean nil.

?	There is some information but no figures. Question marks are used more specifically as follows:
10?	A speculative figure produced in devising the table.
0?	The source reported no evidence, (eg. of under-age recruitment of girls in some cases where women were known to serve.)
?00, ?000	No numbers to hand, but indicating to the correct number of digits, the best guess of the scale of the figure.
com.	committees
Dem.	Democratic
Govt.	Government
Nat.	National
opp.	armed opposition groups
param.	paramilitary force(s)
Rep.	(self-styled) Republic
Sec.	Security

Unless otherwise specified, acronyms are of opposition groups (see Annex 3 for explanations).

Dates: Information given in the Table is "the latest available", usually from 1993 or later. In some cases it pre-dates the most recent political developments. Where the information available relates only to a period before about 1992, it has not been incorporated in this Table.

Most population figures are estimates for 1993 or 1992.

The figure for armed forces excludes reserves.

Recruitment age for volunteers: Where two figures are given, the lower is usually for service in the army, the higher for other services (often the navy). Period of full-time service only recorded, excluding any refresher courses or time in reserves.

Treaties: Indicates if the country is party to the 1949 Geneva Conventions (GC), Additional Protocol I (API), and Additional Protocol II (APII). The Convention on the Rights of the Child, which also contains provisions on recruitment and participation, has been universally ratified, with the exceptions of the Cook Islands, Oman, Somalia, the United Arab Emirates, the USA and Switzerland (ratification in progress).

A	B	C	D	E	F	G	H	I	J	K	L	M
Country	**Population** *(millions)*	**Armed Forces; Paramilitaries and/or Militias; Armed Opposition Groups**	**Child soldiers:**				**Deaths in War or Political Violence 1965-1995** *(per 000 population, approx.)*	**Conscription:** *Yes, No or S (Selective) and Duration of Service (months)*	**Legal Recruitment Ages:**		**Voting age**	**Treaties:** *GC API APII*
			Total number	*As % of force*	*% Girls*	*Lowest age recorded*			*Con-scription*	*Volun-teers*		
Afghanistan	18.1	?00 000	?0000	45	0?	12	50.0	?	?			GC
Albania	3.4	40 000					-	Yes (24/36)	18		18	GC/API-II
Algeria	25.9	150 000					4.0	Yes (6)	19			GC/API-II
FIS		10 000-15 000										
GIA		?000	?00			15						
Andorra	0.06	none	-	-	-	-	-	-	-	-	18	GC
Angola	10.6	128 000	?0000	?		8	50.0	Yes	18		18	GC/API
UNITA		60 000	?000	?		8						
Antigua & Barbuda	0.1	700					-				18	GC/API-II
Argentina	33.5	71 000					0.5	Yes (6/14)	19		18	GC/API-II
Armenia	3.6	20 000	?000	?			2.5	Yes (24)			18	GC/API-II
Australia	17.6	63 200	650	1	16	16	-	No		16	18	GC/API-II
Austria	7.8	52 000	?00	?			-	Yes (6)	18	17	19	GC/API-II
Azerbaijan	7.3	30 000	?000	?			2.5	Yes (24)				GC
Bahamas	0.3	2 600					-	No			18	GC/API-II
Bahrain	0.5	7 200					-	No		18		GC/API-II
Bangladesh	115.1	107 000	?00	?			1.0	No		17	18	GC/API-II
JSS/SB		5 000										
Barbados	0.3	200					-	No			18	GC/API-II
Belarus	10.4	103 000					-	Yes (18)			18	GC/API-II
Belgium	10.0	75 800					-	No		18	18	GC/API-II
Belize	0.2	700					-	No			18	GC/API-II
Benin	5.1	4 300					?	S			18	GC/API-II

A	B	C	D	E	F	G	H	I	J	K	L	M
Country	**Population** *(millions)*	**Armed Forces; Paramilitaries and/or Militias; Armed Opposition Groups**	**Child soldiers:**				**Deaths in War or Political Violence 1965-1995** *(per 000 population, approx.)*	**Conscription:** *Yes, No or S (Selective) and Duration of Service (months)*	**Legal Recruitment Ages:**		**Voting age**	**Treaties:** *GC API APII*
			Total number	*As % of force*	*% Girls*	*Lowest age recorded*			*Con-scription*	*Volun-teers*		
Bhutan	1.4	11 000	700?	5	0	11	-	S (1/3)	18	16	18	GC
Royal Bhutan Police		8 000										
Royal Body Guard		2 000										
militias		>2 200	>200	10?								
Bolivia	7.7	31 500					-	S (12)	21		21	GC/API-II
Bosnia & Herzegovina	3.5	110 000	?000	10?	0?	11	15.0	?				GC/API-II
Rep. Srpska		50 000-80 000	?00	?	0?	14		?		16		
Rep. Herzeg-Bosna		50 000	?00	?	0?			?				
irregulars		?0 000	?00	?	0?	10						
Botswana	1.3	6 100					-	No			21	GC/API-II
Brazil	156.5	296 700					-	Yes (12)	19		16	GC/API-II
Brunei Darussalam	0.3	4 500					-	No				GC/API-II
Bulgaria	8.5	107 000					-	Yes (18)	18		18	GC/API-II
militias		100 000										
Burkina Faso	9.8	7 200					-	No			18	GC/API-II
Burma/Myanmar	44.6	286 000	?0000	10-66		10	0.5	No		18	18	GC
Peoples Police Force		50 000										
Peoples Militia		35 000										
opp.		24 000-40 000	?000	10-40	?	10						
Burundi	5.6	7 200	?00	?	0?	16	35.0	No		16	18	GC/API-II
opp.		?0 000	?000	?	?	12						
Cambodia	9.3	103 000	4 000	4	?	8	100.0	Yes	18	18	18	GC
param./local militias		35 000										
Khmer Rouge		12 000	3 000	25	?	5						

A	B	C	D	E	F	G	H	I	J	K	L	M
Country	**Population** *(millions)*	**Armed Forces; Paramilitaries and/or Militias; Armed Opposition Groups**	**Child soldiers:**				**Deaths in War or Political Violence 1965-1995** *(per 000 population, approx.)*	**Conscription:** *Yes, No or S (Selective) and Duration of Service (months)*	**Legal Recruitment Ages:**		**Voting age**	**Treaties:** *GC API APII*
			Total number	*As % of force*	*% Girls*	*Lowest age recorded*			*Con-scription*	*Volun-teers*		
Cameroon	13.1	8 100					-	No			20	GC/API-II
Canada	28.1	78 000	?00	?	?		-	No		16	18	GC/API-II
Cape Verde	0.3	1 300					-	S			18	GC/API-II
Central African Republic	3.0	3 800					?	S			18	GC/API-II
Chad	6.1	25 000					15.0	S				GC
Chile	13.5	92 000	?00	?			10.0	Yes (24)	18	16	18	GC/API-II
China	1158.0	3 010 000					0.5	S (36/48)	18		18	GC/API-II
paramilitary		12 000 000										
Colombia	34.0	140 000	4 756	3	-	16	2.5	S (12/24)	15		18	GC/API-II
FARC/ELN		10 000–25 000	?000	10–30	?	8						
paramilitary		?000	?000	50?	-	9						
Comoros	0.5	900					-				18	GC/API-II
Congo	2.8	10 800					?	No		18	18	GC/API-II
paramilitary		6 100										
Costa Rica	3.2	none	-	-	-	-	-	No	18		18	GC/API-II
paramilitary		7 500										
Côte d'Ivoire	12.5	7 100					-	S	21		21	GC/API-II
paramilitary		7 800										
Croatia	4.8	110 000	?00	?			2.0	Yes (10)	18	17	18	GC/API-II
Rep. Srpska Krajina		?0 000	?00	?								
Cuba	10.7	176 000	?000	?			-	Yes (36)	17?		16	GC/API
paramilitary		119 000										
Cyprus	0.8	10 000					?	Yes (26)	18		21	GC/API-II
Turkish Cypriot forces		4 000										

A	B	C	D	E	F	G	H	I	J	K	L	M
Country	**Population** *(millions)*	**Armed Forces; Paramilitaries and/or Militias; Armed Opposition Groups**	**Child soldiers:** *Total number*	*As % of force*	*% Girls*	*Lowest age recorded*	**Deaths in War or Political Violence 1965-1995** *(per 000 population, approx.)*	**Conscription:** *Yes, No or S (Selective) and Duration of Service (months)*	**Legal Recruitment Ages:** *Con-scription*	*Volun-teers*	**Voting age**	**Treaties:** *GC API APII*
Czech Republic	10.3	107 000					-	Yes (12)	18		18	GC/API-II
Denmark	5.2	32 800					-	Yes (4/12)	20		18	GC/API-II
local defence force		18 000										
home guard		69 200										
Djibouti	0.7	3 300					?	No			18	GC/API-II
paramilitary		600										
Nat. Sec. Force		1 200										
FRUD		?	?00	?								
Dominica	0.1	none	-	-	-	-	-	-	-	-	18	GC/API-II
Dominican Republic	7.6	22 000					-	Y	18		18	GC/API-II
Ecuador	11.0	58 000	?	?			-	S (12)	19	18	18	GC/API-II
Egypt	57.1	430 000					?	S (36)	18		18	GC/API-II
El Salvador	5.5	30 500	?000	?			15.0	No		16	18	GC/API-II
Equatorial Guinea	0.4	3 100					-	Yes	18		18	GC/API-II
paramilitary		2 000										
Eritrea	3.7	60 000					25.0					
Estonia	1.5	2 500					-	Yes (12)	18		18	GC/API-II
Ethiopia	51.8	100 000					25.0	S (12/18)	18		18	GC/API-II
Fiji	0.8	5 000					-	No			21	GC
Finland	5.0	33 000	?	?			-	Yes (8/11)	18	17	18	GC/API-II
France	57.5	450 100					-	Yes (10)	18/19		18/23	GC/APII
gendarmerie		95 700										
Gabon	1.3	4 800					-	No			21	GC/API-II
paramilitary		2 800										
gendarmerie		2 000										

A	B	C	D	E	F	G	H	I	J	K	L	M
Country	**Population** *(millions)*	**Armed Forces; Paramilitaries and/or Militias; Armed Opposition Groups**	**Child soldiers:**				**Deaths in War or Political Violence 1965-1995** *(per 000 population, approx.)*	**Conscription:** *Yes, No or S (Selective) and Duration of Service (months)*	**Legal Recruitment Ages:**		**Voting age**	**Treaties:** *GC API APII*
			Total number	*As % of force*	*% Girls*	*Lowest age recorded*			*Con-scription*	*Volun-teers*		
Gambia	1.0	200					-	No			21	GC/API-II
armed police force		600										
Georgia	5.5	20 000					1.0	Yes (24)			18	GC/API-II
Germany	79.8	447 000	?00	?			-	Yes (12)	18	17	18	GC/API-II
Ghana	15.6	6 900					1.0	No			18	GC/API-II
Greece	10.3	159 200	?00	?			-	Yes (19/24)	18	16	18	GC/API-II
gendarmerie		26 800										
Grenada	0.1	none	-	-	-	-	-	No			18	GC
Guatemala	9.7	44 600	?000	?	0?	12	15.0	Yes (30)	18	15	18	GC/API-II
PAC militias		800 000	?000	?	0?	11						
URNG		800-1 100	?00	?	?	12						
Guinea	7.4	9 700					-	Yes (24)		19		GC/API-II
Paramilitary		9 600										
Guinea Bissau	1.0	7 200					-	No			18	GC/API-II
paramilitary		2 000										
Guyana	0.7	1 700					-	S			18	GC/API-II
paramilitary		3 500										
Haiti	6.9	7 400					1.5	No			18	GC
Holy See	0.001	-					-	-				GC/API-II
Honduras	5.2	16 800	?000	?		13	-	Yes (24)	18	17	18	GC/API-II
paramilitary		5 000										
guerrilla groups		?	?	?	?	15						
Hong Kong	6.0	none	-	-	-	-	-	-	-	-		
Hungary	10.3	81 000					-	Yes (12)	18		18	GC/API-II
border guard		20 000										

A	B	C	D	E	F	G	H	I	J	K	L	M
Country	**Population** *(millions)*	**Armed Forces; Paramilitaries and/or Militias; Armed Opposition Groups**	**Child soldiers:** *Total number*	*As % of force*	*% Girls*	*Lowest age recorded*	**Deaths in War or Political Violence 1965-1995** *(per 000 population, approx.)*	**Conscription:** *Yes, No or S (Selective) and Duration of Service (months)*	**Legal Recruitment Ages:** *Con-scription*	*Volun-teers*	**Voting age**	**Treaties:** *GC API APII*
Iceland	0.3	none	-	-	-	-	-	-	-	-	18	GC/API-II
India	897.6	1 265 000					-	No			18	GC
opp. (Kashmir)		?	?00	?		12						
Indonesia	188.2	283 000	?000	?			2.5	S (24)		17	17	GC
paramilitary		180 000										
KAMRA militias		1 500 000										
FRETILIN		?	?	?								
Iran	60.8	528 000	?000	?		<15	10.0	Yes (24)	No age limits		16	GC
opp.		?	?00	?								
Iraq	19.5	382 000					50.0	Yes (21/24)	19	18	18	GC
opp. (Kurds)		?	?00	?		10						
Ireland	3.5	13 000	?	?			-	No		17	18	GC
Israel/Occupied Territories	6.5	176 000	?000	?	?		5.0	S (24/36)	18	17	18	GC
Shaba/Hamas, etc.		?	?00	?								
Italy	57.1	354 200					-	Yes (12)	18		18/25	GC/API-II
carabinieri		111 400										
Jamaica	2.5	3 400					-	No			18	GC/API-II
Japan	124.7	238 000					-	No		18	20	GC
Jordan	3.3	99 400					1.0	Yes	18		19	GC/API-II
Kazakhstan	17.2	44 000					-				18	GC/API-II
Kenya	28.1	22 000					0.1	No			18	GC
paramilitary		4 000										
Kiribati	0.1	none	-	-	-	-	-	-	-	-	18	GC
Korea (North)	22.6	1 130 000					-	S			17	GC/API
Korea (South)	44	633 000					-	Yes (30/36)	18		20	GC/API-II

A	B	C	D	E	F	G	H	I	J	K	L	M
Country	**Population** *(millions)*	**Armed Forces; Paramilitaries and/or Militias; Armed Opposition Groups**	**Child soldiers:**				**Deaths in War or Political Violence 1965-1995** *(per 000 population, approx.)*	**Conscription:** *Yes, No or S (Selective) and Duration of Service (months)*	**Legal Recruitment Ages:**		**Voting age**	**Treaties:** *GC API APII*
			Total number	*As % of force*	*% Girls*	*Lowest age recorded*			*Con-scription*	*Volun-teers*		
Kuwait	1.4	13 700					1.0	Yes (24)	18	18	21	GC/API-II
Kyrgyzstan	4.4	12 000					-	No				GC/API-II
Laos	4.5	37 000	?000	?			1.0	Yes (18)	15		18	GC/API-II
Latvia	2.6	4 900					-	Yes (18)			18	GC/API-II
Lebanon	2.8	39 000	?00	?			50.0	Yes	18	18	21	GC
paramilitary		9 000										
SLA		2 500	?00	?								
Palestinian Commandos		?000	?00	?		9						
Lesotho	1.9	2 000					-	No		18	21	GC/API-II
Liberia	2.9	9 000	0?				50.0	-	18			GC/API-II
ECOMOG		15 000-17 000										
various factions		59 000	12 800	25	1?	6						
Libya	4.6	70 000	?000	?			-	Yes (36/48)	18	14	18	GC/API-II
Liechtenstein	0.03	none	-	-	-	-	-	-	-	-	20	GC/API-II
Lithuania	3.8	7 000					-	Yes	19		18	
National Guard		12 500										
Luxembourg	0.4	800	?	?			-	No		17	18	GC/API-II
Macedonia	2.1	20 000					-	Yes (9)			18	GC/API-II
Madagascar	13.4	23 000					-	Yes			18	GC/API-II
Malawi	10.6	10 400					-	No		18	18	GC/API-II
Malaysia	19.1	127 500					-	No			21	GC
Maldives	0.2	none	-	-	-	-	-	-	-	-	21	GC/API-II
Mali	8.6	7 400					?	S			18	GC/API-II
paramilitary		7 800										
Malta	0.4	1 700					-	No			18	GC/API-II

A	B	C	D	E	F	G	H	I	J	K	L	M
Country	**Population** *(millions)*	**Armed Forces; Paramilitaries and/or Militias; Armed Opposition Groups**	**Child soldiers:** *Total number*	*As % of force*	*% Girls*	*Lowest age recorded*	**Deaths in War or Political Violence 1965-1995** *(per 000 population, approx.)*	**Conscription:** *Yes, No or S (Selective) and Duration of Service (months)*	**Legal Recruitment Ages:** *Conscription*	*Volunteers*	**Voting age**	**Treaties:** *GC API APII*
Marshall Islands	0.05	none	-	-	-	-	-	-	-	-	18	
Mauritania	2.2	15 300	?00	?			?	No		16	18	GC/API-II
paramilitary		5 700										
Mauritius	1.1	none	-	-	-	-	-	-	-	-	18	GC/API-II
paramilitary		1 300										
Mexico	90.0	175 000	?000	?			-	Yes (12)	17		18	GC/API
Rural Defence Militia		14 000										
Micronesia	0.1	none	-	-	-	-	-	-	-	-	18	GC/API-II
Moldova	4.4	14 000					-				18	GC/API-II
Monaco	0.03	?					-	No		19	21	GC
Mongolia	2.3	21 000					-	Yes (24)	18		18	GC/API-II
Morocco	26.5	196 000					-	Yes (18)	18		20	GC
Gendarmerie Royale		10 000										
Force Auxilaire		30 000										
Mozambique	15.7	52 000					30.0	S (24)	18		18	GC/API
Namibia	1.4	7 400	?000	?			?	Yes	16		18	GC/API-II
Nauru	0.01	none	-	-	-	-	-	-	-	-	20	
Nepal	19.4	36 000					-	No	18		18	GC
paramilitary		28 000										
Netherlands	15.3	75 000	150	<1	17	16	-	No		16	18	GC/API-II
New Zealand	3.4	10 900	?00	?			-	No		16	18	GC/API-II
Nicaragua	4.3	15 200	?00	?			?	No		17	16	GC
Niger	8.5	5 300					?	S (24)				GC/API-II
Nigeria	88.5	79 000					-	No		18	18	GC/API-II

A	B	C	D	E	F	G	H	I	J	K	L	M
Country	**Population** *(millions)*	**Armed Forces; Paramilitaries and/or Militias; Armed Opposition Groups**	**Child soldiers:**				**Deaths in War or Political Violence 1965-1995** *(per 000 population, approx.)*	**Conscription:** *Yes, No or S (Selective) and Duration of Service (months)*	**Legal Recruitment Ages:**		**Voting age**	**Treaties:** *GC API APII*
			Total number	*As % of force*	*% Girls*	*Lowest age recorded*			*Con-scription*	*Volun-teers*		
Norway	4.3	33 500	?	?			-	Yes	18	17	18	GC/API-II
home guard		80 000										
Oman	2.2	26 500					-	No				GC/API-II
Pakistan	128.0	580 000					0.1	No		18	21	GC
paramilitary		238 000										
Panama	2.6	11 700					-	No			18	GC/API-II
Papua New Guinea	3.9	3 800	?00	?			0.5	No			18	GC
BRA		?	?00	?								
Paraguay	4.6	16 500	10 500	>60	-	14	-	Yes (18/24)	18	18	18	GC/API-II
paramilitary		8 000										
Peru	22.9	115 000	?000	?	?	11	1.0	S (24)	18	16	18	GC/API-II
paramilitary		70 000										
civil defense com.		100 000	?000	?	?	9			15			
Sendero Luminoso		3 000	?00	?		9						
MRTA		500	?	?		12						
Philippines	65	117 000	0?				0.4	Yes	18		18	GC/APII
NPA		7 500-10 000	1 000-2 500?	10-30	30?	10						
MNLF		15 000										
Poland	38.3	296 300					-	Yes (18/24)	18	17	18	GC/API-II
Portugal	10.4	58 300					-	Yes (4)	21		18	GC/API-II
Nat. Republican Guard		20 900										
border guard		8 900										
paramilitary		20 000										
Puerto Rico	3.5	-					-	No				
National Guard		?										

A	B	C	D	E	F	G	H	I	J	K	L	M
Country	**Population** *(millions)*	**Armed Forces; Paramilitaries and/or Militias; Armed Opposition Groups**	**Child soldiers:** *Total number*	*As % of force*	*% Girls*	*Lowest age recorded*	**Deaths in War or Political Violence 1965-1995** *(per 000 population, approx.)*	**Conscription:** *Yes, No or S (Selective) and Duration of Service (months)*	**Legal Recruitment Ages:** *Con-scription*	*Volun-teers*	**Voting age**	**Treaties:** *GC API APII*
Qatar	0.5	7 500					-	No				GC/API
Romania	22.8	201 000					-	Yes (12/18)			18	GC/API-II
Russian Federation	148.0	2 720 000	?00	<1		17	0.2	Yes (18)	18		18	GC/API-II
Chechen separatists		?0 000	?000	?	0?	11						
Rwanda	7.5						>100.0	No		18	18	GC/API-II
(former) Govt. forces		30 000	?	<1								
militias		?0 000	?000	?	0?	15						
FPR		20 000										
Saint Kitts & Nevis	0.04	none	-	-	-	-	-	-	-	-	18	GC/API-II
Saint Lucia	0.1	none	-	-	-	-	-	-	-	-	18	GC/API-II
Saint Vincent & the Grenadines	0.1	none	-	-	-	-	-	-	-	-	18	GC/API-II
San Marino	0.02	?					-	No			18	GC/API-II
São Tomé & Príncipe	0.1	900					-				18	GC
Saudi Arabia	15.3	106 000					-	Yes				GC/API
National Guard		75 000										
frontier force		10 500										
Senegal	1.7	9 700					-	S (24)		18	18	GC/API-II
Seychelles	0.1	1 300					-	Yes (24)			18	GC/API-II
paramilitary		800										
Sierra Leone	4.5	6 200	1 000-1 500	16-24	3	5	2.0	No			18	GC/API-II
RUF		? 000	2 000	?								
Singapore	2.9	55 000					-	Yes (24/30)	18		21	GC
Slovakia	5.3	50 000					-	Yes (12)			18	GC/API-II
Slovenia	2	20 000					-	Yes (7)	18		18	GC/API-II

A	B	C	D	E	F	G	H	I	J	K	L	M
Country	**Population** *(millions)*	**Armed Forces; Paramilitaries and/or Militias; Armed Opposition Groups**	**Child soldiers:** *Total number*	*As % of force*	*% Girls*	*Lowest age recorded*	**Deaths in War or Political Violence 1965-1995** *(per 000 population, approx.)*	**Conscription:** *Yes, No or S (Selective) and Duration of Service (months)*	**Legal Recruitment Ages:** *Conscription*	*Volunteers*	**Voting age**	**Treaties:** *GC API APII*
Solomon Islands	0.3	none	-	-	-	-	-	-	-	-	18	GC/API-II
Somalia	7.1	20 000	?000	?		11	10.0	-			18	GC
South Africa	40.7	68 000	?0000	30?		16	0.5	No		17	18	GC/API-II
ANC/PAC		26 000	50	<1	55	16						
SDUs		>10 000	?0000	>50		11						
SPUs		?	0?									
Spain	40.0	201 000					-	Yes (9)	20		18	GC/API-II
paramilitary		75 000										
Sri Lanka	17.6	110 800					2.0	No		18	18	GC
LTTE		6 000-10 000	?000	?	?	10						
Sudan	25.0	73 000	?000	?		12	40.0					GC
PDF (param.)		40 000	?000	?		7						
SPLA		30 000-50 000	?000	?		11						
Surinam	0.4	1 800					-	No			18	GC/API-II
Swaziland	0.8	2 700					-	S (24)			18	GC/API-II
Sweden	8.6	60 500					-	Yes (7/15)	18		18	GC/API-II
Switzerland	6.9	676 000					-	Yes (4)	20		18	GC/API-II
Syria	13.0	408 000					-	Yes (30)	19		18	GC/API
gendarmerie		8 000										
Taiwan	20.9	360 000				15	-	Yes (24)				
Tajikistan	5.6	3 000	?00	?		?	5.0				18	GC/API-II
Popular Dem. Army			?00	?		?						
Tanzania	25.1	46 000					-	Yes (24)	18		18	GC/API-II
militias		100 000										
Thailand	57.8	280 000					-	Yes (24)			18	GC

A	B	C	D	E	F	G	H	I	J	K	L	M
Country	**Population** *(millions)*	**Armed Forces; Paramilitaries and/or Militias; Armed Opposition Groups**	**Child soldiers:**				**Deaths in War or Political Violence 1965-1995** *(per 000 population, approx.)*	**Conscription:** *Yes, No or S (Selective) and Duration of Service (months)*	**Legal Recruitment Ages:**		**Voting age**	**Treaties:**
			Total number	*As % of force*	*% Girls*	*Lowest age recorded*			*Con-scription*	*Volun-teers*		*GC API APII*
Togo	3.8	5 300					?	S (24)			18	GC/API-II
paramilitary		800										
Tonga	0.1	300					-	No			21	GC
Trinidad & Tobago	1.2	2 700					-	No			18	GC
paramilitary		?										
Tunisia	8.5	35 000					-	Yes	20	18	20	GC/API-II
Public Order Brigade		3.5										
National Guard		10										
Turkey	60.0	560 000	0?				0.2	Yes (18)	18		18	GC
village guards		70 000	0?									
PKK		10 000-12 000	>3 000	25-30	>10	7						
Turkmenistan	3.9	28 000					-	Yes (18)			18	GC/API-II
Tuvalu	0.01	none	-	-	-	-	-	-	-	-	18	GC
Uganda	16.6	70 000					10.0	No			18	GC/API-II
LRA		5 000?	>2 000	?	>30	11						
Ukraine	52.3	230 000					-	Yes (18)	18		18	GC/API-II
United Arab Emirates	1.9	55 000					-	No				GC/API-II
United Kingdom	58.1	293 500	4 500	2	11	15	-	No		16	18	GC
IRA/UDF		?	?00	?	?	13						
USA	256.6	1 730 000	?000	<1		17	-	No	18	17	18	GC
Uruguay	3.2	24 700					?	No	18		18	GC/API-II
Uzbekistan	21	40 000					-	Yes (18)			18	GC/API-II
Vanuatu	0.2	300					-	No			18	GC/API-II
Venezuela	20.7	53 000					-	S (24)	18		18	GC
National Guard		22 000										

A	B	C	D	E	F	G	H	I	J	K	L	M
Country	**Population** *(millions)*	**Armed Forces; Paramilitaries and/or Militias; Armed Opposition Groups**	**Child soldiers:**				**Deaths in War or Political Violence 1965-1995** *(per 000 population, approx.)*	**Conscription:** *Yes, No or S (Selective) and Duration of Service (months)*	**Legal Recruitment Ages:**		**Voting age**	**Treaties:**
			Total number	*As % of force*	*% Girls*	*Lowest age recorded*			*Con-scription*	*Volun-teers*		*GC API APII*
Vietnam	70.9	857 000					20.0	Yes (36)	18		18	GC/API
Western Samoa	0.2	none	-	-	-	-	-	-	-	-	21	GC/API-II
Yemen	12.5	65 000	?00	?			2.0	Yes (24/36)			18	GC/API-II
paramilitary		20 000										
Yugoslavia *(Serbia & Montenegro)*	10.6	135 000	?00	?			-	Yes (12)	18	17	18	GC/API-II
Zaire	41.1	29 700					0.5	Yes	18		18	GC/API
gendarmerie		25 000										
civil guard		18 000										
Zambia	7.8	21 600					-	No			18	GC/API-II
Zimbabwe	10.7	48 200					-	S (12)			18	GC/API-II

FOOTNOTES TO TABLE 2

Afghanistan

C No reliable figures of the current strength of the Government forces or of the various factions are available. Estimates given here are therefore for all parties to the conflict taken together. Principal armed opposition groups are Hezb-i-Islami, Hezb-i-Wahdat, Jumbush-i Milli-ye Islami, and Taliban.

G Source: Defence for Children International (DCI), International Children's Rights Monitor, 1994 (Vol. 11): No. 4/1995 (Vol. 12): No 1, p. 12.

I In 1978, according to the case study, the conscription age was 22 years, but this was progressively lowered. By 1991, according to the Submission by War Resisters International to the Human Rights Committee, June 1994 ("Issues of Conscience and Military Service"), the age was 15. "Military service lasted for four years, followed by a two-year gap and a further two years." The current situation is unclear.

Algeria

G Source: Le drame Algérien: un peuple en otage, Reporters sans frontières, Editions de la Découverte, Paris, 1996.

Angola

G Source: Human Rights Watch, Children In Combat 8,1(G), (1996)

Armenia

H Figure for Armenia and Azerbaijan together.

I Source: Guinness.

Australia

D Details on recruitment of under-18s provided by the Australian Defence Force. (Letter dated 1st May 1996).

Austria

L This age has to be attained the year before the election year.

Azerbaijan

H Figure for Armenia and Azerbaijan together.

I Source: Guinness

Belgium

I Conscription (formerly eight months) abolished in 1994.

Bhutan

B Government figures tend to be only some 60% of those from outside sources; the Government contests the size of the population of Nepali origin.

C Source: case study. Since 1990, the army and police have allegedly doubled in numbers and the militia groups have been formed. For the army total, a statement to the National Assembly by the Chief of Army in August 1995 is quoted. The total strength of the militias is not known; the figure of 2,200 represents only those detachments specifically mentioned in the case study.

E The case study quotes testimonies that one detachment of 200 soldiers contained three very young soldiers "(not more than 15 years)", and that another of 400 contained 15 who were under 16 years. It also quotes a testimony that in one group of 300 militia recruits there were 25 to 30 "very young boys". If typical, these observations would imply the approximate percentages given here, from which may be extrapolated speculative totals. Given the ages mentioned and the unknown total size of the militias, these estimates may well be conservative.

F According to the case study only the Royal Bhutan Police currently takes female recruits, and no claim is made of child recruitment into that force.

I Source: Guinness.

K Source: case study, quoting government officials and a National Assembly resolution.

L Source: Goodwin-Gill & Cohn.

Bolivia

L If married, 18 years.

Bosnia & Herzegovina

B Excluding approximately 1 million refugees currently elsewhere.

D An estimate for the former Yugoslavia is 3,000–4,000, the overwhelming majority in Bosnia and Croatia.

G Source for Bosnian army: Barnen och vi,1995, No 6. (Swedish magazine from Rädda Barnen)

I It is believed that all three communities operate some form of conscription, but there is no precise information.

K The official minimum recruitment age into the army of Republika Srpska is 17, but recruits are taken from the beginning of the year in which they will attain this age (case study).

Burkina Faso

I Source: Guinness.

Burma/Myanmar

C The lower total of opposition forces is derived from SIPRI's estimate of the strength of the two largest groups, the MTA (Mong Tai Army) and the KNU (Karen National Union), the higher from estimates quoted in the case study of the 1989 strengths of some 30 armed opposition groups and sub-groups.

E Range of estimates quoted in case study for different regiments/units and by different informants.

L Source: Goodwin-Gill & Cohn

Cambodia

C Estimate for militias based on SIPRI estimate of total government forces at 130,000 to 140,000. Khmer Rouge strength as quoted in the case study; SIPRI gives 6,000 to 15,000.

I The case study quotes a 1989 constitutional provision, but in practice recruitment is local and haphazard. Guinness describes the system in the words "military service is selective and varies in length".

Central African Republic

I Source: Guinness.

Colombia

C Information on child soldiers and opposition and paramilitary groups comes from a report from Defensoría de Pueblo (Human Rights Ombudsman) of May 1996. The Government denies responsibility for paramilitary groups. The case study also documents the existence of popular militias, engaged in "self-defence of the population ... at present, some sectors of the militias may have relations with guerilla groups."

E The CNG (Guerilla coordination council) admits to 7–10% child soldiers. Sources in the armed forces and former guerillas quote total figures of around 30%; up to 80% in urban guerilla units and lower in rural areas.

H The report from Defensoria del Pueblo does not mention guerillas below the age of 13, but the case study guotes two references to children of 8 years within the guerilla groups.

I Source: Guinness.

J Source: Report from Defensorìa del Pueblo, May 1996. Conscription occurs at the age of majority (18), with the exception for high school students, who can be enrolled when they leave high school (from the age of 15). In 1995–96, 22% of all recruits were under 18.

Costa Rica

J Constitutional provision exists for the establishment of conscripted armed forces, should that become necessary.

Croatia

D An estimate for the former Yugoslavia is 3,000–4,000, the overwhelming majority in Bosnia and Croatia.

Cyprus

C Excluding mainland Turkish forces (about 30,000) in the occupied zone.

Czech Republic

I Recently reduced from 18 months.

L Source: Goodwin-Gill & Cohn.

Denmark

J According to Guinness "conscripts are selected at the age of 18–19 and are called up at about the age of 20".

Djibouti

B Including over 100,000 refugees, principally from Somalia.

D Source: Dorothea E. Woods, Geneva.

Dominican Republic

L If married, the voting age can be lower.

El Salvador

I Abolished 1992 (Guinness).

Equatorial Guinea

L Source: Goodwin-Gill & Cohn.

Eritrea

C This is the estimated size of the EPLF forces before independence.

H Including Ethiopia. It is not clear to what extent this figure includes famine deaths.

Ethiopia

C Information relates to pre-1992 period. Despite the collapse of the Derg regime in 1992 and the subsequent secession of Eritrea, a number of armed opposition groups remain active in Ethiopia, but no up-to-date figures (including figures on child soldiers) are to hand.

H Including Eritrea. It is not clear to what extent this figure includes famine deaths.

France

I The government in 1996 announced its intention of abolishing conscription.

L The voting ages refer to the National Assembly and the Senate, respectively.

Gambia

I "Military service is normally voluntary, but is compulsory in certain circumstances".

L Source: Goodwin-Gill & Cohn.

Grenada:

C Except a small paramilitary Special Service Unit.

Guatemala

C The estimate for PAC militias is quoted in the case study, which also estimates the total armed forces strength as 62,000. The Guatemalan government announced in August this year that all the PACs (Civil Defence Patrols) will be disarmed and demobilised by 15 November 1996.

D During the months of May and June 1995, the Attorney General's Office for Human Rights (Procuradoria de Derechos Humanos) received 596 complaints from young men who claimed to have been recruited by force, and obtained the release of 333 of them, of whom 148 were under 18 (presumably at the time of release). If indeed these cases referred only to a two-month period, they could reflect anything between 1,000 and 4,000 forced under-age recruitments per annum.

I Source: case study. Guinness says "two years or more".

Guinea

I Source: Guinness.

Guyana

I Selective military service in the paramilitary.

Haiti

I Source: Guinness.

Holy See

C There are 100 Swiss Guards.

Honduras

A Honduras has at no time in the recent past been classified as involved in an armed conflict, under any definition. Even so, the case study documents the existence of opposition guerilla groups which recruit minors.

D The case study quotes estimates from the annual reports of CODEH (The Honduran Committee for Civil Rights) that there were some 3,690 cases of forced recruitment, mainly of minors, in the five-year period 1988–1992.

I Source: case study (Guinness quotes 8 months). Soldiers have to claim release; it is not automatic.

Hong Kong

C The United Kingdom is responsible for defence.

Iceland

C Although a member of NATO, Iceland has no defence forces of its own except a 130-man coastguard. 3,000 United States troops are stationed in Iceland.

India

D Source: Dorothea E. Woods, Geneva.

G Source: Dorothea E. Woods, Geneva.

H The rate in Kashmir is approximately 2 per thousand of the Kashmiri population.

I "There is no military service but it is the legal duty of every citizen to perform national service if called upon."

Indonesia

D Source: Dorothea E. Woods, Geneva.

H The number of deaths associated with the occupation of East Timor are, according to common estimates, over 200 per thousand of the East Timorese population.

L If married at the time of registration on electorial lists, the voting age can be lower.

Iran

D Source: Dorothea E. Woods, Geneva.

G Source: Dorothea E. Woods, Geneva.

Iraq

D Source: Dorothea E. Woods, Geneva.

G News report, Swedish Radio (Morgonekot), 16th Sept. 1996.

Israel/Occupied Territories

D Source: Dorothea E. Woods, Geneva.

I "Compulsory for all Jewish Israeli men and women... voluntary for Christian and Arab Israelis." (Guinness).

Italy

L The voting ages refer to the Chamber of Deputies and the Senate, respectively.

Japan

K But 15 in military schools (stated in UN Working Group on Optional Protocol to Convention on the Rights of the Child, February 1996).

Kenya

C Based on figures given by Guinness for the separate services, which are not consistent with the total quoted.

Korea (North)

L Source: Goodwin-Gill & Cohn.

Kuwait

L Refers to men only.

Latvia

I Source: Guinness.

Lebanon

D The case study documents also the participation of children in the various factional militias during the 1975–90 civil wars, specifically quoting testimonies from former child soldiers held on criminal charges. (In December 1994, five years after the supposed end of hostilities, 646 children were in detention on charges involving arms).

I Source: Action Pour une Alternative Civile.

Liberia

C The source for ECOMOG (the West African States' Peacekeeping Mission) is SIPRI. Other figures, including those for the nominal government forces, come from the case study, quoting the First Progress Report of the Technical Committee of the Task Force for Disarmament, Demobilisation, and Reintegration, August 1995.

D Source: case study, as above. Although they represent the most informed estimates available, they should be treated with caution. Children are known to drift in and out of the conflict and the different factions, making an accurate count impossible.

F Two of a sample of 319 demobilised children were girls.

I Government forces have effectively ceased to exist.

J Source: case study.

Lithuania

M Lithuania has ratified Geneva Conventions I–III.

Macedonia

C Increased from 10,400 during 1993 and 1994.

I Source: Guinness.

Malawi

B Including about 1 million refugees from Mozambique.

Maldives

C The police force is responsible for national security.

Marshall Islands

C The United States is responsible for defence.

Mexico

I Eligibility for military service decided by lottery.

Micronesia

C The United States is responsible for defence.

Monaco

C "There is a small palace guard." (Guinness).

Mozambique

I Both men and women liable for service.

Nauru

C Australia is responsible for defence.

Netherlands

D Information on recruitment provided by the Netherlands Mission to the United Nations, Geneva. (Letter dated 14th June 1996).

I Conscription (at age 19) was abolished 1994.

Nicaragua

I Abolished 1990 (Guinness).

Nigeria

L Source: Goodwin-Gill & Cohn.

Panama

C Mainly paramilitary police.

Papua New Guinea

D Source: Private letter.

H Mainly in Bougainville.

Paraguay

I The case study quotes a 12-month period of service.

Peru

A Sendero Luminoso = Shining Path.

C Shining Path forces were considerably depleted by the offensive by the security forces in 1992. At their peak they are believed to have forcibly recruited several thousand children from indigenous communities in areas under their control.

J Source: case study. Men and women liable.

K Source: case study. Men only, and only with the father's written authority.

Philippines

C The figure for MNLF is a 1989 estimate. Source: ITN Factbook, Guild Publishing, London, 1990.

D Based on reported proportions.
E Varying between units.
F If interviews reported in the case study were representative.

Puerto Rico
C Service in US forces applies.

Russian Federation
H The rate in Chechnya may be over 40 per thousand of the Chechen population.

Rwanda
A Militias include Interhamwe and Impuzamugambi.
B Figures for population and armed forces in Rwanda are as at the eve of the massacres and refugee outflow of April 1994.
L Source: Goodwin-Gill & Cohn.

Saint Lucia
C A paramilitary unit is attached to the police.

Saint Vincent & the Grenadines
C A paramilitary unit is attached to the police.

Sâo Tomé and Príncipe
C Paramilitary gendarmerie.

Seychelles
I Source: Guinness.

Sierra Leone
A Source on child soldiers: Case study reported in McCallin, M., The reintegration of young ex-combatants into civilian life, ILO, 1995.
F Based on 370 demobilised children analysed in the ILO case study.
L Source: Goodwin-Gill & Cohn.

Somalia
B Plus about a million refugees currently in neighbouring countries.
C There is no effective government. The figures given relate to the Aideed and Mahdi factions.
G Source: Reuter, Oct 17, 1995.
H Estimates including indirect casualties are more than three times as high.
L Source: Goodwin-Gill & Cohn.

South Africa
A SDUs are (unofficial) Self-Defence Units allied to ANC, and SPUs (unofficial) Self-Protection Units allied to Inkatha.
C Strengths as in 1993. Source for non-governmental forces: case study.
D ANC/PAC: Provisional figure as at April 1994 of admission (with parental consent) of former combatants into unified national force.
E ANC/PAC: Recruitment into the armed wings of the liberation movement was

reportedly strictly controlled. Many more children took part in unstructured violent opposition, and/or fled the country to camps controlled by the liberation movements. SPUs: The case study was able to investigate these only in the Gauteng province, where the Inkatha supporters are almost exclusively immigrant workers. The situation in Natal, Inkatha's home territory, may well be different.

I Abolished 1993. Formerly (white) conscription on leaving school – minimum age 16.

Spain

I Abolition currently proposed.

Sri Lanka

C Estimate of government forces is dated 1993. SIPRI estimate that by 1994 the total was 126,000.

Sudan

A Source on child soldiers: Human Rights Watch, Children in Combat 8,1(G), (1996).

H This figure may include a large number of indirect casualties.

Surinam

I Abolished 1993.

Taiwan

G Source: Far Eastern Economic Review, Oct. 5, 1995.

Tajikistan

D Source: Private correspondance.

Tanzania

I Source: Guinness.

Tonga

C National Police Force.

Trinidad & Tobago

C Plus paramilitary police force.

Turkey

C The case study quotes PKK sources as estimating their strength at 15,000 (which they hoped to double) in 1994.

J Source: case study. Goodwin-Gill & Cohn quote 20.

Turkmenistan

I Source: Guinness.

Tuvalu

C The 32 member police force is responsible for security.

Uganda

F Based on 183 former abductees surveyed in January 1995.

L Source: Goodwin-Gill & Cohn.

United Kingdom

D Details on recruitment of under-18s provided by UK Mission to the United Nations, Geneva (Letter of 11th June, 1996).

K In practice recruits are accepted at 15 years and 8 months.

H The rate in Northern Ireland is approximately 2 per thousand of the Northern Ireland population.

United States of America

I Provision exists for selective call-up in time of war.

Uruguay

I Source: Guinness.

Uzbekistan

I Source: Guinness.

Vanuatu

C Paramilitaries.

Western Samoa

C New Zealand is responsible for defence.

Yemen

H Estimates of deaths in the North/South war of 1994 range from 7,000 to 60,000. This entry is thus more than usually speculative.

I Source: Guinness. Conscription applies "in theory. Length of service is 3 years in the former North Yemen, 2 years in the former South Yemen."

Yugoslavia

D An estimate for the former Yugoslavia is 3,000–4,000, the overwhelming majority in Bosnia and Croatia.

Zaire

I "In theory" (Guinness).

L Source: Goodwin-Gill & Cohn.

Parade during the Armed Forces Day in Guatemala.
Photo: James Nachtwey/Magnum.

CHAPTER 2

Recruitment

It is definitively a problem of political ethics, the recruitment laws themselves establish the age but if states themselves allow the armed forces not to respect the law we shall never get anywhere.[38]

It is no accident that such a heavy emphasis in this book is placed on recruitment. Recruitment is the defining moment. Without the recruitment of children, there would be no child soldiers. Many child soldiers are invisible because they grow into, and become demobilised as, adult soldiers.

In drafting the questionnaires, reference was made to compulsory, forced, voluntary and induced[39] recruitment, and the writers of the case studies took pains to relate their answers to these categories. What emerges, however, is that the areas of overlap between the categories are more striking than the differences. In practice there would appear to be a continuum between forced and voluntary recruitment, with the true facts in many cases being unascertainable.

CONSCRIPTION

Of all the categories, that of compulsory recruitment, or conscription, is the most distinct. This consists of the legal obligation of citizens falling into the specified category to perform a stated period of obligatory military

38 Dr Ramon Custodio, President of the Committee for the Defence of Human Rights in Honduras, quoted in the case study for Honduras (translated from the Spanish)

39 Induced recruitment was defined as those situations where there is no proof of direct threat or intimidation but the circumstances suggest that the enrolment is not voluntary.

service. Conscription is practised in many countries in all regions of the world. In some it has been introduced as a response to conflict, in others it is a standing arrangement. In some countries, what are referred to in this book as "child soldiers", that is, those under 18, may be legally conscripted. In the majority of cases, however, military service is required of males (and sometimes females) of 18 years and over (see Table 2 for the current position).

It is notable that none of the case studies mentions a possibility of exemption from military service on grounds of conscience, or the availability of alternative civilian service for those with a conscientious objection to military service.

By its nature, conscription is a governmental prerogative. In one instance, the armed opposition group issued its own "Compulsory Military Service Law", aimed at members of the ethnic minority of military age (18–25 years).[40] However, in reality this was simply a way of encouraging volunteers and of legitimising forced recruitment to pre-empt conscription into the government armed forces. In fairness, it should be observed that the same applies to much that is done by governments in the name of conscription. For, unfortunately, the existence of conscription with a legal minimum age of eighteen is no guarantee against the recruitment of child soldiers into government forces. The main reasons for this, which will be discussed below, are: lack of documentation, voluntary enlistment for compulsory military service, "quota" enlistment, and what may be termed legal, systematic and infrastructural inadequacies.

Lack of Documentation: The simplest, and the most intractable, problem is lack of birth or identity records. The absence of these leads to genuine problems in establishing the age of potential recruits, as well as serving to facilitate under-age recruitment:

> *It is fairly usual for age to be estimated based on stature, psycho-motor development, and, in the case of girls, the appearance of breasts.*[41]

[40] Case study for Turkey

[41] Case study for Mozambique (translated from the Portuguese)

Even where documentation does exist, there are circumstances where it commonly overstates the true age of the child:

> *In the case in which the elder brother dies, of disease or accident, at an early age, to avoid the painfully long process of issuing state identification cards for the younger brother, some families have preferred to transfer the identity of the deceased elder to the younger. As a result, early recruitment has taken place.*[42]

Or it may even represent a *posthoc* rationalisation:

> *For the majority ... especially for those living in rural areas precision in the year, let alone in the month and day of birth, is an unattainable abstraction. Hence age is determined not by having regard to any official record but by having regard to what the person says his/her age is, physical appearances, or even by what the person seeking the information thinks the age is. In relation to children's involvement in armed conflicts and armed forces, this situation means that those below 18 are officially recorded as above 18 simply because the recruitment of children is officially not acceptable and there is no way of ascertaining the actual age of the persons concerned.*[43]

This uncertainty, credible in itself, may readily be used either as an excuse for failure to make such checks of age as are possible, or to mask conscious and systematic under-age recruitment. Those forcibly recruited or volunteering are encouraged or forced to state that they are 18 in order to ensure apparent conformity with national legislation or international norms. When questions are asked, the authorities blame the recruits for misstating their ages. Alternatively, "informal lists" may be kept of under-age soldiers rather than including them in official records.[44]

42 Case study for Turkey

43 Case study for Ethiopia

44 Case studies for Burma/Myanmar, El Salvador, Ethiopia, Guatemala, Mozambique and Paraguay; see also Human Rights Watch/Africa, Children's Rights Project: *Children of Sudan* (September 1995), p 26

Voluntary Enlistment for Compulsory Military Service: Sometimes, faced with the inevitability of military service, recruits find it more convenient to enlist before the official age. In many cases voluntary enlistment is permitted at a lower age than that of conscription. Sometimes there is specific provision for voluntary induction in order to complete compulsory military service at an earlier and more convenient time. These loopholes may be exploited to mask systematic recruitment of children, with under-age conscripts being claimed as "volunteers". In other circumstances, instances have been documented of under-age candidates presenting themselves with forged documents for recruitment.[45]

"Quota" Enlistment: Not all conscription systems rely upon an individual call-up. Indeed, such a practice can only function where there is relatively accurate and comprehensive documentation of the population. Often it is the number rather than the identities of the conscripts which are pre-determined. One method of filling the quota is the conscription lottery which takes different forms in different countries. One case study described a system by which all those deemed eligible on the basis of height ("A rifle was used as the measure of appropriate height!") had to draw out a piece of paper, some of which were blank and some said "join the army". However, the apparently random nature of the process was belied by evidence that the lottery was rigged in advance so that certain families or individuals were targeted, in particular those who had fallen foul of the recruiter.[46]

If those entrusted with producing the recruits are unable, for whatever reason, to come up with the requisite number of the official recruitment age, they may be tempted, or forced, to make up the shortfall with under-age recruits:

> *Personnel of the armed forces entrusted with the task of recruitment were officially under a duty to exclude children from recruitment. They did not live up to expectations, however, due to several factors. The*

[45] Case studies for El Salvador and Turkey

[46] Case study for Cambodia

major reason for this is that the main task of these personnel is to obtain a pre-set number of recruits within a specified period of time. Since not so many volunteered to join the army, the recruiters take as many of those as present themselves unless their recruitment is blatantly not conforming to the criteria for recruitment including the age requirement.[47]

These agents may be detachments of the military. On the other hand, the task of finding recruits may be devolved to local militias/civil defence forces, or the headmen or other authorities of the local community:

> *The district/sub-district Administrator ... instructs the village headman to provide a specific number of people from his village for "voluntary" recruitment, which he is obliged to meet. Thus several (mostly) young people from the district are rounded up in district headquarters for initial medical and other check-ups by authorised Armed force personnel.*[48]

Such indirect methods of conscription are particularly prone to corruption and substitution, both of which increase the number of children recruited. The selection may in practice be influenced less by age than by the inability to pay bribes, whether to the recruiter or to a potential substitute, having had a disagreement with the village leader or for example, a younger brother may be put forward in order to allow his older brother to finish his education.

A quota system will often coexist with the requirement that only a proportion of the eligible population need perform military service, although neither of these conditions relies upon the other. Where the two coincide there are implications not only for the likelihood of the recruitment of child soldiers but also on who is vulnerable to such recruitment, as discussed in Chapter 3. In such circumstances, there is in practice little difference between conscription and forced recruitment. Indeed, armed opposition groups sometimes themselves obtain recruits by imposing

[47] Case study for Ethiopia
[48] Case study for Bhutan

quotas upon communities within their area of operation; sometimes the recruits go willingly, sometimes not.[49]

Legal, Systematic and Infrastructural Inadequacies: Where the conscription system is fundamentally flawed, rather than simply subject to abuse at the margins, conscription frequently merges into forced recruitment, particularly if it can be done away from public scrutiny, because the conflict is taking place in a remote area, for example, or if disenfranchised sections of the community can be targeted.

The constitutional provision for conscription may not be translated by legislation into appropriate procedures, so that in practice implementation is at best random. Alternatively, the military may simply ignore the prescribed procedures. Even where laws are in place to punish those engaging in under-age recruitment, these may never be applied, and thus be ineffective.[50]

The system or infrastructure may be inadequate to produce the required number of recruits particularly if, as a result of conflict, large numbers are required over a prolonged period of time, or if the mechanisms for enforcement are so inadequate that a high proportion of those eligible escape service and alternative recruits have to be found. Evasion is particularly common where the conflict, or the military, are unpopular, or where casualties are known to be high.[51] If salaries are inadequate and desertions high, "ghost soldiers" may be maintained on the lists in order to provide the commanders with surplus salaries while children are recruited to fill the ranks.[52] On the other hand, as will be discussed below, there may be a deliberate policy of the military to target certain groups and the political structures may lack the strength, or the desire, to challenge it.

49 Case studies for Burma/Myanmar, Cambodia, Lebanon and Liberia

50 Case studies for El Salvador, Honduras and Paraguay

51 Case studies for Afghanistan, Cambodia, El Salvador, Ethiopia, Guatemala and Paraguay; see also Human Rights Watch/Africa, Children's Rights Project: *Children of Sudan* (September 1995), p 55

52 Case study for Cambodia

Causes: Forced recruitment into government armed forces may be a response to an immediate shortfall in manpower,[53] as discussed above. However, in some countries there is a long tradition of forced recruitment.

In addition to the generally disenfranchised, there are particular groups which are especially subject to forced recruitment. These may be specific ethnic, racial or religious groups, such as the indigenous population, because they are perceived as a threat to the government. This threat may be as a recruiting ground for armed opposition groups, or their general supporters or potential supporters of it, or merely as the "sea" in which the opposition groups "swim" (in Mao's phrase). Such recruitment frequently forms part of a general campaign of intimidation, repression or disempowerment and disruption.[54]

As already mentioned, forced recruitment into armed opposition groups may sometimes be justified as a form of conscription, a requirement that all members of the ethnic group contribute to the armed struggle.[55] Otherwise much the same reasons apply as in the case of government armed forces: the group is under pressure and needs additional soldiers, which is particularly likely if the conflict is prolonged, and is compounded if the group is unpopular and therefore the number of volunteers is inadequate, or if there is a high level of desertions.[56] Whereas the government can, at least theoretically, use legal means to draw on the entire population,

53 Case studies for Afghanistan, Bhutan, Burma/Myanmar, Cambodia, Guatemala, Lebanon, Nicaragua, Paraguay and the former Yugoslavia; see also Human Rights Watch/Africa, Children's Rights Project: *Children of Sudan* (September 1995), p 55

54 Case studies for Burma/Myanmar, El Salvador, Ethiopia, Guatemala, Mozambique, Nicaragua, Paraguay and Peru

55 Case studies for Burundi, Lebanon and Turkey

56 Case studies for Afghanistan, Burma/Myanmar, Cambodia, Lebanon, Liberia, Mozambique, Turkey and Uganda; see also Human Rights Watch/Africa, Children's Rights Project: *Children of Sudan* (September 1995), p 77

armed opposition groups can recruit actively only in the territory they control or have access to, thus at best they have a limited pool of manpower. One case study referred to the problem being compounded by a high level of migratory labour, causing a local shortage of able-bodied men.[57]

Methods of Forced Recruitment: The most common method of forced recruitment by both government and armed opposition groups is that described in the case study for Ethiopia:

> *An extreme version of forcible recruitment is the press ganging. This is known as "Afesa" in Amharic. A group of armed militia, police, or party cadres would roam the streets and marketplaces, picking up any individual or rounding up any groups they came across. Alternatively, they would surround an area and force every man and boy to sit down or stand up against a wall, using a threat of opening fire. All those eligible would then be forced on to a truck and driven away. Young men were recruited while playing football, on side streets and alleyways, going to school or market places or attending religious festivals. Teenage boys who worked in the informal sector selling cigarettes, matches, sweets, chewing gum and lottery tickets were a particular target.*[58]

Very similar accounts occur in different cultures:

> *The forced recruitment took place in the poor suburbs, at the football grounds, movie theatres, at the bus stops or in front of big schools, etc; places where young men and boys of the lower classes of society moved in major quantities. Army trucks also went to poor distant villages in the countryside and swept up the streets picking up the young men.*[59]

> *Other children were recruited at school gates, market places and other areas where people gather together.*[60]

> *"We were leaving school at the end of the day and the ... soldiers sur-*

[57] Case study for Mozambique
[58] Case study for Ethiopia
[59] Case study for El Salvador
[60] Case study for Mozambique (translated from the Portuguese)

rounded the school ... There were 40 or 50 of us all leaving together, and we were all arrested. We were all 15, 16, 17 years old, and we were all afraid of the soldiers. We were students, we looked like students, because we were all wearing our blue shirts and green longyis. Our teachers all ran away in fear. Everything was in chaos ... We were all terrified, but we couldn't even call out to them to let us go and that we were under 18, because we were so scared ... I didn't know what was going on and they didn't explain anything to us."[61]

In rural areas, both government and armed opposition groups use similar methods. The army or group enters villages and small towns, killing people, abducting children, looting and burning homes. Children are abducted from their homes at night or during the day from home or school or from working in the fields. In almost all instances they are taken by force or with threats and intimidation:[62]

"They take the boys and men as porters, rape some of the women, steal the chickens and pigs, steal the rice and burn what we have in the fields."[63]

Those who resist would be cut with pangas. Quite a number of victims had their lips and ears chopped off in macabre rituals.[64]

Various methods may be used to try to escape such recruitment; the cost of failure can be high:

If there was an "afesa" in a neighbourhood, local women would patrol near the area and warn men and boys to stay away, or give them "gabis" to hide under to disguise themselves as women. If all failed, people would run for hiding places into nearby houses, granaries, ward-

61 Case study for Burma/Myanmar

62 Case studies for Burma/Myanmar, Cambodia, Colombia, El Salvador, Liberia, Peru and Uganda; see also Human Rights Watch/Africa, Children's Rights Project: *Children of Sudan* (September 1995), p 56, and Amnesty International, "Sierra Leone: Human rights abuses in a war against civilians", 13 September 1995, pp 19 and 23

63 Case study for Burma/Myanmar

64 Case study for Uganda

robes, ceilings, cattle fodder, empty barrels, etc. There were also numerous instances of people trying to resist or escape press-ganging being summarily killed.[65]

At the margins, "press-ganging" by government forces can be described as an enforcement of conscription legislation. There is evidence that in some situations there is enough respect for age limits that those who can produce documentary proof that they are under-age are released, even if this tends to be associated with a preference for concentrating upon those groups who are unlikely to have such proof.[66] More frequently, age is at best a matter of indifference to those recruiting in this way; they are often satisfied with obviously under-age recruits:

> *Many children whose age has been mentioned clearly in their national ID cards as less than 18 years were taken to [a] special military commission where the military officers amended their age to meet the criteria of military service. In this way they were sending children aged less than 14 years to the armed forces.*[67]

There is reason to believe that children are disproportionately vulnerable to forced recruitment procedures. Recruiters prefer to concentrate on those who can resist least effectively, which includes children.

Sometimes, however, the evidence is that the recruitment of children is a deliberate policy rather than a consequence of other factors. The frequent references to recruitment in or near schools[68] point to this conclusion. Recruitment from schools is one instance in which those who are the most disadvantaged may not be the target, depending on how widespread the availability of schooling is. One specific instance was reported of a focus on national educational institutes to produce more educated recruits in order to combat urban armed opposition groups.[69]

65 Case study for Ethiopia

66 Case studies for Colombia, Ethiopia and Guatemala

67 Case study for Afghanistan

68 Case studies for Afghanistan, Bhutan, Burma/Myanmar, El Salvador, Ethiopia and Mozambique

69 Case study for El Salvador

As previously mentioned, a frequent expedient of both government and armed opposition groups is to conduct their recruitment by proxy. Recruitment is entrusted to local authorities or militias, who have to fill a quota with the penalty for failure being to be sent to prison or on active service themselves.[70] This frequently degenerates into forced recruitment:

In an over-zealousness to fulfil the quota and to avoid the much dreaded active military service, the militias and military commissariats resorted to forcible and arbitrary recruitment. Trying to avoid the ill-will of their neighbourhoods, territorial militias in many cases picked up recruits from among strangers in market places, places of worship ... and on streets. They were not, however, sparing of their neighbours also, and sent letters of recruitment, and followed these with night time searches.[71]

However, even under age, specific individuals may be sought and, if they are not found, pressure may be brought to bear on their families. A vivid instance is provided by the following testimony of a former child soldier:

"I was 16 years old ... while my younger brother was 14 years old ... One afternoon a neighbour came to our home and told my father that the people's militia have a plan to visit our home during the night for the purpose of recruiting either myself or my younger brother ... my father called myself and my younger brother and told us that he will be taking us to a far away hiding place ... On the third or the fourth day, my younger brother urged me that he should go back home and find out what has happened ... When he reached home, he found out that our house was searched at the said night and the people's militia trying to find either of us, have detained both of our parents as hostages. He went straight to the police station, where they were detained, submitted himself for recruitment, and demanded the release of our parents. Upon their release, my parents came to take me home because they have got an assurance that not more than one person will be recruited from a single family. When I learned the fate of my brother I felt like crying. It

[70] Case studies for Burma/Myanmar and Ethiopia

[71] Case study for Ethiopia

felt as if I was guilty of betraying my younger brother. What is more, I could not bear the sight of my mother and sisters crying, and the thought of my younger brother being recruited to spare me, made me restless. In short I hated myself ... In the evening I somehow managed to sneak out of the house and went to the recruiting officials with the intention of substituting my younger brother. They did not show him to me then. It was several months later, at the training camp, that I learned my brother was not released and was in fact placed in a different training centre. He died a year later while in combat ..."[72]

The various forms of local militia (which are known in different places by such titles as civil defence patrols, self-defence committees, peasant patrols or temporary village guards) are an under-documented phenomenon whose role in recruiting children for themselves and for regular government armed forces merits further research.[73] Characteristically justified by governments because of the security situation in conflict areas, persons recruited into them are often under threat of otherwise being accused of being guerillas or sympathisers, with dire consequences for the individuals or the village concerned. Children, in particular, often serve on behalf of their parents, other economically active adults, or old or infirm people who are incapable of carrying out the duties required. Some militias may be genuine self-defence groups organised by young people from the local community in the face of inadequate protection from, or violations by, the government, but there are questions about possible links to one or other party. Alternatively, what began as local "peaceful, democratic and autonomous" organisations can be taken over by the government and armed, or permitted to carry arms, as part of its anti-subversion strategy. Where such forces are well paid and are created in an area of high unemployment, this attracts volunteers, but also creates employment which is entirely dependent on the continuation of the conflict.[74]

72 Case study for Ethiopia

73 Case studies for Cambodia, Colombia, Guatemala, Mozambique, Peru and Turkey; see also Human Rights Watch/Africa, Children's Rights Project: *Children of Sudan* (September 1995), pp 31, 57

74 Case studies for Colombia, Guatemala, Mozambique, Peru and Turkey

Recruitment by intimidation rather than by crude compulsion is certainly not a practice confined to government forces. Even where armed opposition groups may genuinely believe that they are using persuasion rather than force, the children may feel that they have been recruited against their will.[75] Also, when a heavily armed group enters a village and makes speeches calling for volunteers, how voluntary is the recruitment? What are the perceived prospects of resisting?

On the other hand, if supplying volunteers to the armed opposition group bears the risk of retribution by government forces, there may be a systematic practice of reporting them as having been abducted. This may also be seen as affording some measure of protection to the individuals themselves should they fall into government hands:

> *What was common knowledge even to local officials at that time was, however, that although the [opposition] recruitment activities were reported as kidnapping by the national press ... this was mainly the result of the fear shared by the villagers of possible [government] retribution to suspected voluntary participation ... It was definitely in the interest of the family and the person taken by the [opposition group] to be reported as kidnapped in any case. First, he or she may have been truly abducted, but families would fear [government] interception of the specific guerilla group and attempt to protect their relatives by emphasizing the point of having been kidnapped. Then there was the possibility that the individual would volunteer to join the [opposition group] but this would leave the family with the burden of explaining his sudden disappearance to local officials – especially if the said person is obliged to serve the [government] army. Even before being trained by the guerillas they could be intercepted by [government] forces and caught. There was thus fear that if the potential recruits returned to their villages in one way or the other, before or after training, they could be punished by the authorities if suspected of voluntary recruitment. As a result of this, most voluntary recruitment was also reported by the villagers themselves to local authorities as kidnap. There were also cases ... during*

75 Case study for El Salvador; see also Amnesty International, "Sierra Leone: Human rights abuses in a war against civilians" (13 September 1995), p 23

which the individuals joined the [opposition group] of their own free will and after initial military training returned to their settlements to conduct undercover military activities. Such people were explained off as having managed to "escape" from the [opposition group] or "released" at one stage. Another frequent explanation behind kidnap claims, as reported by families and militants concerned, appeared to be concern on the part of the volunteer or the family that the individual would not fit rural war conditions and would be turned down half-way by the guerilla unit escorting them ... Yet there are many recorded incidents of [opposition group] abductions in the history of the conflict ...[76]

In short, the truth about individual cases may be impossible to ascertain in the context of the conflicting pressures.

Less directly, some governments and armed opposition groups are known to use the creation of educational establishments and institutions for street children and orphans as providing training grounds for the military,"warehousing" them against future need.[77] Those enrolled in such establishments may include children captured from "enemy" positions or abandoned in front line villages:

> ... *orphans between 0–12 years, were taken by [the opposition group] into refugee camps and daycare centres, where military training was sometimes secretly given by [opposition] combatants ... many of those children joined [the opposition group] later, at the age of about 12 years.*[78]

Such institutions may be an attempt to respond to a genuine problem of what to do with unaccompanied children. However, they also play on the lack of security and educational opportunity for children. At best they undermine the child's free choice as to whether or not to volunteer. At worst they are simply a disguised form of forced recruitment.

[76] Case study for Turkey

[77] Case studies for Burma/Myanmar, Ethiopia, Mozambique, Peru and Sri Lanka; see also Africa Watch: *Angola Civilians Devastated by Fifteen Year War* (5 February 1991), p 12; and Human Rights Watch/Africa, Children's Rights Project: *Children of Sudan* (September 1995), pp 5, 56, 77–83

[78] Case study for El Salvador

VOLUNTARY RECRUITMENT

Although many of these children and adolescents have entered voluntarily into participation in this type of activity, the whole concept of what is "voluntary" needs to be called into question. Various indirect coercive mechanisms have been used on these minors, such as physical protection, the stimulation of machismo, the symbology of concepts such as defense of the fatherland, the heroic nature of enlistment, revenge and adventure. Or through economic, cultural and social considerations such as belief in the justice of the cause, social pressure, provision of food etc. Therefore, the dividing line between voluntary and forced participation is very imprecise and ambiguous.[79]

The previous section considered the various methods by which recruits, particularly children, are selected against their will by armed forces or groups or their agents. This section considers why some children make a positive choice to join, or are encouraged or constrained to volunteer by force of circumstances, or where the choice is made on the child's behalf by the family. As much of the material illustrates, too great an implication of freedom of choice should not be associated with the term "voluntary" in this context. Nevertheless, certain motives and causes for voluntary recruitment can be identified and may be broadly grouped as cultural, in search of protection, ideological, and economic and social.

Cultural Reasons for Volunteering: In some situations, participation in military or warlike activities is glorified. Children are raised to revere military leaders of the past, and to look on military induction as a sign of manhood. At least to certain sections of opinion, the younger the age at which the child is involved in military activity, the more laudable. From the point of view of the child, the military life or the glamour and prestige of a military uniform may be attractions. Others may be persuaded to join by their parents in order to keep them off the streets, for political or patriotic reasons, or through a belief that military discipline is beneficial for the child.[80]

79 Case study for Peru (translated from the Spanish)

80 Case studies for Burundi, Burma/Myanmar, Colombia, El Salvador, Ethiopia, Guatemala, Liberia and Nicaragua

Khmer Rouge fighters in "The Children's Army" in Galaw, Cambodia.
Photo: Pica Pressphoto.

In some cases the young and especially the illiterates believed they gained prestige by belonging to an 'armed faction' with resulting power and authority which are unparalleled with what other children can achieve ... many become drunk with power and often abuse it.[81]

Children may succumb to the seductive lure of being able to command others, and of no longer being prey to their dictates ... To be a soldier is to occupy a position of great honour and self-sacrifice. The emotional pull of such prestige should not be underestimated.[82]

The charm of being part of an army, the weapons, the uniforms and the recognition of permanent role models were some of the reasons for enlisting ... The commandantes, for the children and young people, were the male figures which exercised the greatest attraction ...[83]

The need to distinguish themselves, to prove themselves as men, to become popular among their peers and neighbours was the motivation of the child soldiers mostly from urban settings; they would walk around the city with guns and in camouflage uniforms, perceiving themselves as important and older.[84]

Sometimes the incentive to become involved in the conflict comes from peer pressure rather than from the wider society. Some join simply because their friends have joined, particularly in urban areas and garrison towns. Children join armed opposition groups or participate in the conflict for the "adventure", "attracted by the sheer fun of belonging", or in order to become "famous and admired".[85]

Value systems which endorse bearing arms as a mark of masculinity can also draw, or push, youngsters into armed opposition groups, especially where linked to a tradition of "blood revenge", which requires even children

81 Case study for Liberia

82 Case study for Burma/Myanmar

83 Case study for Colombia (translated from the Spanish)

84 Case study for the former Yugoslavia

85 Case study for Colombia, El Salvador, Ethiopia, the Intifada, Liberia, Sri Lanka and the former Yugoslavia; see also case study for Sierra Leone in M. McCallin: *The reintegration of young ex-combatants into civilian life* (ILO, Geneva, 1995)

to avenge the deaths of members of the extended family or even more distant kinsmen.[86]

Traditional values may also allow or encourage the use of existing youth groups, such as scouts, or the establishment of youth movements specifically to instil military values, commitment to the cause, and sometimes to provide familiarity with weapons, for example how to dismantle and clean them.[87]

In many cases, sections of the population, for example, those living in certain areas, or belonging to certain ethnic groups may feel a general, passive loyalty to the armed opposition group, or the wider movement, rather than to the government. This, quite distinct from any passionate involvement in the cause or tradition of blood revenge, may result in a sense of obligation simply to replace, rather than to revenge, a father, brother or other relative already lost in the struggle.[88]

In Search of Protection: Sometimes it is a desire for revenge that motivates the volunteer:[89]

> *... these youngsters may wish to avenge the genocide of their tribes, the torture and murder of their parents and the multiple rape (in the case of females) to which they were subjected.*[90]

> *Torture, abuse killing, deprivation and humiliation are all experiences which can create desires for revenge. Others include bombings, harassment, detention, and intimidation by security forces.*[91]

However, it is remarkable how rarely in the case studies the children themselves cite revenge as a motive for volunteering.

[86] Case studies for Chechnya and Ethiopia
[87] Case studies for Afghanistan, Lebanon and Sri Lanka
[88] Case study for El Salvador
[89] Case studies for Burma/Myanmar, the Intifada and Liberia
[90] Case study for Liberia
[91] Case study for Sri Lanka

We came across only one soldier who explicitly gave the desire for revenge as his primary motivation.[92]

Blood revenge, like military tradition, seems to be a reason why families and communities "volunteer" their children. For the children themselves, far more often than revenge the motive seems to be their sense of vulnerability: enrolment in the government armed forces may be seen as a means to protect themselves and their families from harassment, whether by armed opposition groups or by the government forces themselves.[93]

What emerges as the single major reason why children volunteer for armed opposition groups is their own personal experience of harassment by government armed forces, including torture, loss of home or family members, enforced exile within the country or out of the country:

... most of the young individuals join [the opposition group] owing mainly to [a] reaction to extensive human rights violations which they have personally witnessed from [the] security forces and most specifically, [the] systematic scorched earth policy ...[94]

Many of the youngest to join [the opposition] ranks were children, who had lost both parents in the war, either through death or exile, and had no-one else to take care of them. Many had seen their parents captured and/or tortured, even assassinated by army soldiers, their houses burnt, properties destroyed or robbed. They joined to look for protection.[95]

Young [ethnic] males often joined the combatants in order to avoid arbitrary detention. [Government] security forces in [the territory] often detained young males between the ages of 14 and 18 as potential combatants, in order to prevent them from joining the rebel forces.[96]

[92] Case study for Cambodia

[93] Case studies for Burma/Myanmar, Chechnya, El Salvador, the Intifada, Liberia, South Africa, Turkey and Uganda; see also case study for Sierra Leone in M. McCallin: *The reintegration of young ex-combatants into civilian life* (ILO, Geneva, 1995)

[94] Case study for Turkey

[95] Case study for El Salvador

[96] Case study for Chechnya

This would also appear to be the case where the involvement of the children in the conflict is spontaneous rather than as a result of recruitment:

> *They could not avoid the negative psychological consequences of harassment, humiliations, injuries, detentions, torture, or the death and injury of friends and relatives, the sealing or destruction of their homes, the nightly raids into homes in the refugee camps, seizing sleeping suspects and beating family members, particularly fathers – normally at the top of the family hierarchy – in full view of everyone else.*[97]

Others join because they feel it is better to die fighting than defenceless:[98]

> *It could become a much more attractive option to being helpless and afraid at home.*[99]

Little evidence was presented to suggest the converse: that mistreatment of civilians or recruits by armed opposition groups increases the number of volunteers entering government armed forces, merely that this reduces the number of volunteers into the group itself. This may be the result of an effective conscription system, making volunteering for government forces unnecessary, or may in part reflect the fact that armed opposition groups usually operate in limited areas and sporadically and therefore it is easier to avoid their activities by temporary or permanent relocation. On the other hand, the only legitimacy which can be claimed by such groups comes from popular support, and they are often aware of the need to keep on at least reasonable terms with the local population and sometimes make considerable efforts to cultivate them. Clearly, where an armed opposition group treats the civilian population and its own recruits well, it increases the number of volunteers entering its own ranks.[100]

97 Case study for the Intifada
98 Case study for El Salvador
99 Case study for Sri Lanka
100 Case studies for El Salvador and the Philippines

Ideological Reasons for Volunteering:

> *Some feel that children are attracted by a stark black and white vision of the world offered by the [armed opposition group]. However, there is information that some young recruits ran away from home to join the movement for the pleasure of riding a motorbike or tractor.*[101]

Some children volunteer for armed opposition groups because they believe in what they are fighting for. This may be synonymous with the stated aims of the group: the holy war, the fight for freedom, freedom to practise their religion, the right to occupy their ancestral lands, ethnic liberation, political liberty. It may, on the other hand, be a general desire for social justice and opposition to poverty, corruption and militarisation, which need not imply a wholehearted endorsement of the ideology of the group's leaders, or stem from a sense of alienation – what other purpose is there in living? The children's commitment to the opposition cause may have been instilled in them throughout their upbringing and be reinforced by an idealisation of the culture of violence.[102]

Those who die for the cause may be considered as martyrs. Children and young people who participate are deemed to be "martyrs" or "honoured fighters", and the community values their participation and is proud of their achievements (even if their own families may be ambivalent, worrying for their safety). The cult of martyrdom plays a particular role in relation to recruitment and training of adolescents – particularly girls – for suicide attacks.[103]

In only one case study[104] is any evidence given of such loyalty or ideologically-based commitment to government forces, but it must be remembered that most of the case studies cover internal conflicts which to a greater or lesser extent pit governments against at least some of their own people. It is reasonable to assume that the situation would be different in the case of international conflicts.

[101] Case study for Sri Lanka

[102] Case studies for Afghanistan, Burma/Myanmar, Chechnya, Colombia, Ethiopia, Honduras, the Intifada, Lebanon, the Philippines, South Africa and Sri Lanka

[103] Case studies for Lebanon and Sri Lanka

[104] Case study for the former Yugoslavia

In some instances the armed opposition group has the support of the community and there is a high level of community involvement directly or indirectly: cultivating and distributing food for the combatants, caring for the wounded, informing about the movements of the army, hiding guerillas and so on. In these cases the communities generally approve of the participation of children and even support it. In any case, since in poor rural families, children, sometimes from as young as four, are actively involved in domestic and agricultural work, they are unlikely to remain outside a conflict which the community supports. Similarly, where 16 year olds are expected to work, and are treated as adults, the community also expects them to be active participants in armed conflicts.[105] The whole family may in fact be participating in the armed opposition group and be proud of their children's participation in the "holy war defending their native land and ideology".[106]

The armed opposition group may put its case by way of speeches calling for volunteers or use other methods of persuasion or indoctrination:[107]

> *... the guerillas arrived, killed a cow for the meal, there was a fiesta with the peasants – coffee-pickers – they stayed for several days and talked with the people.*[108]

> *[an opposition] recruitment unit ... would enter an unguarded hamlet or village, gather all the villagers in a convenient area, propagate on behalf of their organisation and [its aims] and then ask for the local youth to join them.*[109]

Sometimes recruitment has an even stronger ideological element:

> *The procedure for recruitment begins with the [opposition] regular eyeing individuals in the communities in the rural areas, who have made*

[105] Case studies for Afghanistan, Chechnya, Colombia and El Salvador

[106] Case study for Afghanistan

[107] Case studies for Colombia, Guatemala and Turkey; see also Human Rights Watch/Africa, Children's Rights Project: *Children of Sudan* (September 1995), p 85

[108] Case study for Colombia (translated from the Spanish)

[109] Case study for Turkey

an impression on them. Usually, these would be individuals who are open to their cause. The [opposition] regular then approaches the potential recruit and starts an ongoing discussion on the [opposition] – its objectives, vision, rules. Once s/he agrees to join the [opposition] s/he undergoes a six months to one year trial period.[110]

Economic and Social Reasons for Volunteering:

For the refugees, internally displaced, homeless, orphaned and fearful, joining an armed group would appear an attractive option. Inadequacies in education and lack of hope in the future could make the [armed opposition group] an attractive proposition to a youngster. The [armed opposition group] could appear a disagreeable but known entity as opposed to an unknown, feared alternative of displacement and homelessness. Sometimes parents may also encourage children to join out of desperation when there is hunger and poverty in the family.[111]

In a very large number of cases the motivation for volunteering may be to find some means of survival or support, sometimes at the most basic level of feeding the family.[112] The alternative to enlistment may be unemployment.[113] Where it is traditional to have weapons at home and children are familiar with their use, the military option may be the preferred route for solving economic and other problems. This may lead job-seekers, particularly in areas of high unemployment where the army is the major employer, to conceal their real age in order to be recruited.[114]

However, it may be the family, rather than the individual child, who bows to economic pressure, especially where the army pays a percentage of

[110] Case study for the Philippines

[111] Case study for Sri Lanka

[112] Case studies for Afghanistan and Cambodia; see also case study for Sierra Leone in M. McCallin: *The reintegration of young ex-combatants into civilian life* (ILO, Geneva, 1995)

[113] Case studies for Afghanistan, Burundi, Burma/Myanmar, Cambodia, Chechnya, Colombia, El Salvador, Ethiopia, Lebanon, Liberia, Nicaragua, the Philippines, South Africa and Sri Lanka

[114] Case studies for Chechnya and Lebanon

a minor soldier's wages directly to the family. Similarly, armed opposition groups sometimes offer wages or personal allowances or various forms of subsidy for family members, including provision of food, transport or medicines; in one case there is also the possibility of access to work permits in an adjoining country.[115] Sometimes the economic inducement is not of pay, but vaguer promises of riches, or the knowledge that

> *the guns in their hands allow them to extract food forcibly from the civilian population.*[116]
>
> *In situations of deprivation, the gun can also become an entry point to food and survival.*[117]

In some cases, the economic motivation is for more than pure survival. The army has traditionally been one of the few routes of upward mobility in some societies (one of the few positions in society where power can be exercised over others through a post that is neither inherited nor attained through merit), and those poor but better educated children who therefore have prospects of becoming officers and earning a much higher salary may stay beyond the minimum period of service. Involvement in armed opposition groups may hold out to volunteers the prospect of becoming "a commander",[118] or may be a reaction to the frustrations of "educated, unemployed youth".[119]

Children who volunteer for economic reasons are particularly prone to deception. The benefits which they anticipate may not be forthcoming, with inadequate food and clothing.[120] Some specific examples are of children who were led to believe that by joining they would learn a trade and thus be able to earn their living, or were offered paid work with food and lodging clearing barricades from the streets of the capital but were sent to combat units.[121]

[115] Case studies for Cambodia, Guatemala, Lebanon, the Philippines and Sri Lanka
[116] Case study for Liberia
[117] Case study for Sri Lanka
[118] Case study for Burma/Myanmar, Colombia, El Salvador and Turkey
[119] Case study for Sri Lanka
[120] Case study for Burma/Myanmar
[121] Case study for Nicaragua

Sometimes volunteering provides a refuge which is not basically economic. One case study reports an interview with a boy who joined government armed forces in order to escape criminal charges[122]; another that young criminals join "to have strong support of armed groups for robbery, looting and other misconducts."[123] Girls may volunteer for armed opposition groups in order to escape from a marriage or an imminent marriage,[124] or conversely be encouraged by their parents to join because of poor marriage prospects.[125]

122 Case study for Burma/Myanmar
123 Case study for Afghanistan
124 Case study for Ethiopia
125 Case study for Sri Lanka

CHAPTER 3

Who are the Child Soldiers?

Gathering and presenting facts on the topic of child combatants is constrained by the fact that reliable sources of information do not wish to be quoted for fear of ... reprisal. In addition, all reports, however carefully formulated, tend to be dismissed as propaganda by the group criticised.[126]

Where do the child soldiers come from? Irrespective of the method of recruitment, the overwhelming impression from the case studies is that the answer is the same. Firstly, from the poor or otherwise disadvantaged sections of society. Secondly, from the actual conflict zones themselves and thirdly, over and above the overlap with the other categories, from those with disrupted or non-existent family backgrounds.

THE POOR AND DISADVANTAGED

As is shown in Chapter 2, in practice, conscription frequently shades into forced recruitment. Even where this is not the case, however, the incidence is often uneven. The better-off families may send their children abroad, at least ostensibly, for education. While out of the country they will not be liable for military service and they may take pains not to return until the danger is over. Indeed, sometimes the whole family may relocate abroad.[127] The resulting loss to the society of some of its wealthier and better-educated elements may be permanent.

[126] Case study for Sri Lanka
[127] Case studies for Afghanistan, Lebanon and South Africa

Within the country, corruption may be more or less institutionalised as a means of avoiding conscription:

> *It must be made clear that the compulsory nature of military service does not, in practice, apply to adolescents from the middle and upper sectors of society who, by means of money or influence, manage to have themselves declared "unfit" or simply "not selected" in the selection procedures for recruits.*[128]

Or the means of evasion may be indirect: one case study[129] records that it is possible to purchase a military service record, the essential documentary proof that the required service has been undertaken. By contrast, members of various groups who are legally exempt (for example, the only sons of widows or separated mothers, heads of households, or members of indigenous groups) often have difficulty in producing documentary evidence to establish their status. Where conscription is organised on a quota basis, the scope for "buying out" is obviously greater. One case study[130] reported an extreme institutionalisation of this, which was termed the "conscription lottery". Under this system, the names of those eligible for conscription were drawn by lot but, for a fixed price, the family of the person whose name was drawn could pay the army to draw again.

As far as forced recruitment is concerned, even if within recruiting raids individuals are seized at random, the direction of the raids themselves is far from random. Overwhelmingly they target gatherings of the poorer, more disadvantaged groups in society. Nor is this just because such groups tend to cluster together in places where they can easily be rounded up. Government recruiters typically target those who are perceived as a threat: directly (potential recruits to armed opposition groups) or indirectly (supporters or potential supporters of opposition groups). These may be specific ethnic, racial, indigenous or religious groups; they may (as discussed below) be the inhabitants of particular areas; or they may simply be the poorest and most disadvantaged sections of society. Recruiters naturally

128 Case study for Peru (translated from the Spanish)

129 Case study for Colombia

130 Case study for Burma/Myanmar

prefer to concentrate on those who can mount the least effective resistance or challenge, which means the most disenfranchised groups. Over and above these considerations, recruitment may form one weapon in a general campaign of intimidation, repression, or general disempowerment and disruption. There is a consistent pattern that the more influential classes in society and particularly the wealthier parts of urban areas are immune from such recruitment.[131]

Once seized by a "press gang", the recruit has little chance of escape. In some cases there is enough respect for age limits that those who can produce documentary proof of age may be released. However, the typical victim of forced recruitment is unlikely to be able to do so. Often the production of proof depends on the family managing to trace the recruit. However, recruits are typically posted as far as possible from their home area[132] which not only breaks the link with the local community and makes it harder to desert but also makes it more difficult for families to trace the recruits, and may make the search prohibitively expensive. In addition to the physical and financial difficulties entailed, the anguish of the family searching for the child who has disappeared in this way should not be overlooked.

It may be possible to evade recruitment. One case study reports:

> *As people riding in cars were usually safe [from press ganging], employers, friends and relatives with cars would pick up men and boys from school, university or place of work when they heard that there was a danger of "afesa" [press gang].*[133]

Whether or not this is intentional, the fact that those in cars are safe once more helps to shield the relatively prosperous.

There is sometimes the possibility of challenging under-age recruitment through publicity or legal channels; some of the rare instances when chal-

[131] Case studies for Burma/Myanmar, Cambodia, El Salvador, Ethiopia, Guatemala, Mozambique, Nicaragua, Paraguay and Peru

[132] Case studies for Afghanistan, Bhutan, Burma/Myanmar, Colombia, Guatemala, Honduras and Nicaragua

[133] Case study for Ethiopia

lenges have been successfully mounted are discussed in Chapter 8. However, to mount such challenges may require a degree of organisation, education and money.

Finally, and more consistently, there is the possibility of buying oneself out:

> *The main strategy for keeping children out of the army was to bribe the recruiting units or local commanders or village chiefs, depending on who was in charge of the recruitment or conscription. Paying the army not to take a child or adult was common practice as was the practice of paying a poor family or individual to take one's place.*[134]
>
> *The sons of the well-off families did not run the risk of being recruited even in peacetime, and less at the time of the war. If a son of an upper- or middle-class family was taken by mistake in ... the forced recruitment, he was quickly released when the parents showed up to demand him or to pay a bribe for his release.*[135]
>
> *Our parents had no idea what happened to us. They weren't told anything and neither were we. Some people had money to pay off the ... officers, but most couldn't.*[136]

Sometimes buying out may take a disguised form. One case study[137] records that when families do manage to trace illegally recruited children they may have to pay for documents to prove the child's age and be charged for the food, lodging and clothes of the under-age recruit while he was in the armed forces.

All these factors operate in the same direction. Children from the more prosperous, privileged classes suffer far less risk of recruitment into government armed forces, and if by chance they are picked up they are more likely to be released. Thus they rarely become child soldiers.

Nor is the focus on the disadvantaged unique to government recruiters:

[134] Case study for Cambodia
[135] Case study for El Salvador
[136] Case study for Burma/Myanmar
[137] Case study for Guatemala

It has been reported that most of the child combatants are school dropouts and from poorer families. The elite and their children are accorded favoured treatment in return for complicity.[138]

As the fighting among armed forces became intensified and the rebel forces sustained losses in the process, their leaders imposed a quota for new recruits who in most cases comprised street and other disadvantaged children to augment their manpower.[139]

What of volunteers? It is precisely the poor and disadvantaged groups who are most at risk from forced recruitment who also have the strongest economic and social incentives to volunteer. The difference is that military service to the volunteer tends to be seen as a route out of destitution rather than a further aspect or consequence of it. Even more than poverty, educational deprivation is the hallmark of the child volunteer. Limited education or educational opportunities are frequently mentioned in the case studies as a reason for volunteering.[140] Volunteers may be those who

"despair of education and see the armed forces as an adventuresome opportunity to escape from the boredom and tedious routine of formal education".[141]

In other circumstances they may have been attracted by promises of education, study grants, or overseas travel.[142] Lack of education is even more characteristic of volunteers than of forced recruits.

The outcome of all this can be summed up by quoting one analysis of

[138] Case study for Sri Lanka

[139] Case study for Liberia

[140] Case studies for Afghanistan, Cambodia, Chechnya, Colombia, the Intifada, Lebanon, Mozambique, Nicaragua, the Philippines, South Africa and the former Yugoslavia; see also case study for Sierra Leone in M. McCallin: *The reintegration of young ex-combatants into civilian life* (ILO, Geneva, 1995)

[141] Case study for Ethiopia

[142] Case studies for Afghanistan, Chechnya, Colombia, the Intifada, Lebanon, Mozambique, Nicaragua, the Philippines, South Africa and the former Yugoslavia; see also Human Rights Watch/Africa, Children's Rights Project: *Children of Sudan* (September 1995), pp 77–83

the background of demobilised child soldiers, forced recruits and volunteers:

> *... the boys mainly came from poor peasant families in isolated rural areas or from the conflict zones. The poverty of their homes bordered on destitution to judge by the clothing, type of house, environmental conditions and educational level of the families. The child soldiers from urban areas came from homes where the head of the family was a woman; they were the sons of cooks, fruitsellers, small traders. These families had numerous children and they were obviously poor, judging by the materials of which the house was made, their clothing and the areas they lived in ...*[143]

A uniquely disadvantaged sub-group in society, and hence a particularly fruitful source of child soldiers, are the inhabitants of camps for refugees or internally displaced persons.[144] In particular, displaced groups are likely to regroup themselves according to their religion or ethnicity, forming ready recruiting grounds for armed opposition groups if these themselves are so based – the degree of recruitment being correlated with the perceived level of threat from other groups. Furthermore, their general problems are often compounded by inadequate economic provision.[145]

> *Returnees from the border camps and other landless or internally displaced persons are the most vulnerable people in [the country] and it is from these families that boys leave to join up.*[146]

THE INHABITANTS OF THE CONFLICT ZONES

Where a conflict has persisted over many years, the children of the conflict zone will inevitably be among the poorer and most deprived members of

[143] Case study for Nicaragua (translated from the Spanish)

[144] Case study for Cambodia; see also Human Rights Watch/Africa, Arms Project: *Angola: Arms Trade and Violations of the Laws of War Since the 1992 Elections*, (NY: Human Rights Watch, 1994) p 87, and Human Rights Watch/Africa, Children's Rights Project: *Children of Sudan* (September 1995), pp 4, 56, 60–61

[145] Case study for Lebanon

[146] Case study for Cambodia

society and, therefore, particularly prone to recruitment. The local economy is affected by the conflict; educational provision suffers and the wealthier part of the population may well have relocated or at least sent its children to a safer part of the country[147] thereby adding to the economic decline of the region. Also in the conflict zones an unusually high proportion of households will through death and injury have ended up with children as the main income earners.[148] Not only is this likely to be correlated with extreme destitution but, with the economy in tatters, child breadwinners are likely to look to the only growth industry – the armed forces or groups.

However, the population of the conflict zones are at risk over and above their poverty. From the point of view of government armed forces, they are particularly likely to be targeted for recruitment as actual or potential supporters of armed opposition groups. If the armed opposition group practices forced recruitment, it is the population of the conflict areas to whom they have access and who are therefore at risk. It can happen that the same village is obliged to supply a quota of recruits to both the government and the opposition.[149] Children from the same village can find themselves fighting on opposing sides:

> *(One child soldier) tells how he and the other adolescents of his village ... were in the thick of the action all along the demarcation line and in some cases crossed that [same] line to go to their village for the weekend, returning for the working week to their places on the demarcation line, opposite youngsters (from the same region) who they sometimes knew (on the other side).*[150]

In the conflict zones too, all the factors which predispose children towards volunteering are present in acute form, especially the lack of alternative opportunities and the tendency to view the conflict, and participation in it,

147 Case studies for Afghanistan, El Salvador, Lebanon, Liberia, South Africa and Sri Lanka

148 Case study for Afghanistan

149 Case study for Burma/Myanmar

150 Case study for Lebanon (translated from the French)

Chechen boys stand guard at a Chechen check-point west of Grozny 1995.
Photo: Oleg Popov, Reuter/Pressens Bild.

as the normal way of life (also illustrated in the above quotation). This includes situations where family members live in the army camp with the soldiers and the children therefore consider it natural for them to join up too.[151] Above all, it is in the conflict zones that children are likely to have the direct personal experience of abuses by government armed forces identified in Chapter 2 as the main reason for their volunteering for armed opposition groups. The victims of forced resettlements in the conflict zones are particularly prone to choose the alternative of seeking security and stability in membership of an armed opposition group.[152]

The combination of factors in the conflict zones makes the involvement in the conflict more widespread throughout the population of such zones, including children. However, it seems likely that they specifically tend to increase the recruitment of children and of steadily younger children, in particular because there is competition for the available manpower which is diminishing as the conflict continues. Perceiving the danger that children may be recruited into armed opposition groups, government forces attempt to pre-empt this, which is counter-productive, since it tends to increase volunteering into the armed opposition group.[153] Such volunteering is primarily occasioned by harassment or a sense of resignation to the inevitability of being forced to take a part in the conflict and therefore a desire to have some choice in the matter of sides and some belief in the cause for which they are fighting. However, it is also strongly influenced by the perception that treatment of and conditions for recruits in the armed opposition groups are better than in the government forces. At other times in the conflict zones, the armed forces or armed opposition groups may for humanitarian reasons pick up unaccompanied children who, however, end up joining the fighting ranks.[154]

The way in which the conflict becomes an integral part of the lives of some children is shown by the references to their changing sides, either as

[151] Case study for Cambodia

[152] Case studies for Burma/Myanmar and Ethiopia

[153] Case studies for Afghanistan, Burma/Myanmar and El Salvador

[154] Case study for Liberia; see also case study for Uganda in M. McCallin: *The re-integration of young ex-combatants into civilian life* (ILO, Geneva, 1995)

a result of defection or of capture in battle.[155] When this is a case of defection, the motive would usually come under the heading of "ideological"; at other times it may be simply that the children "have got accustomed to the gun culture and do not have skills for any other type of job"[156] or that they may feel the need for protection: "Some of these children may have committed atrocities ... and fear reprisals."[157]

SEPARATED CHILDREN

No group in society is as vulnerable as children separated from their families for whatever reason. Inevitably the conflict zones contain a particularly high proportion of children who have lost or are separated from their families. All the evidence in the case studies suggests that these groups, and other children whose family background is unstable or disrupted – whose fathers have been recruited, detained or killed in the conflict leaving the mother as the head of household, illegitimate children, children of divorced or separated parents and of single-parent families in general – are even more prone than their peers to become child soldiers.[158]

There are various reasons for this. One is that the family provides a measure of physical protection and assistance in strategies for avoiding recruitment. Children with no families have noone to send them away into safety. Accompaniment by adults is sometimes an effective protection against forced recruitment. By contrast, it has been reported that unaccompanied children even from solid family backgrounds, and even when running errands for the family, have been picked up and designated as street children.[159] (Although sometimes whole families are kidnapped to

[155] Case studies for Afghanistan, Burma/Myanmar, Cambodia, El Salvador, Liberia, Mozambique and Uganda

[156] Case study for Uganda

[157] Case study for Uganda

[158] Case studies for Burundi, Burma/Myanmar, Cambodia, Chechnya, Colombia, El Salvador, Ethiopia, the Intifada, Lebanon, Liberia, the Philippines, Turkey and the former Yugoslavia; case study for Uganda in M. McCallin: *The reintegration of young ex-combatants into civilian life* (ILO, Geneva, 1995)

[159] Case studies for Burma/Myanmar and Ethiopia; see also Human Rights Watch/Africa, Children's Rights Project: *Children of Sudan* (September 1995), p 56

the armed opposition base, where the children are given military training.[160])

Unaccompanied children are an attractive target for a recruiting party because on the simple physical level they will be unable to resist as effectively as adults. Once forcibly recruited, such children are unlikely to have anyone to make enquiries, let alone to mount any form of challenge or to pay a bribe. Then again, the family sometimes – though obviously not always – acts as a repository of alternative values:

> *Educated families/parents saw the military struggle as barbaric and uncivilised and therefore did everything to ensure the non-involvement of their children. Some families/parents because of their religious background ... did not allow their children to participate in it. While some parents, because of their experiences [from] the devastating effect of war, stopped their children from participating in the conflict.*[161]

Without families as sources of discouragement, children are more at the mercy of militarist cultures, peer pressure and the inability to conceive of life outside the confines of the conflict. They are also more vulnerable to trickery such as false accusations of theft, leaving the choice of army or prison.[162] Obviously, family pressure is likely to weigh more heavily with the child, the more stable and cohesive the family background itself is.

Finally, it is in the search for a substitute or replacement family that children in these categories are particularly liable to join or attach themselves to armed forces or armed opposition groups:

> *For those without a family, the militia acts as a large family with a leader and brothers in arms.*
>
> *The testimony of former child soldiers in prison reveals why they became involved in the militia: [boy] 16, is illiterate and ran away from home because of on-going differences with his father. [Boy] 15, could not go to school as his father could not afford it (having another child handi-*

160 Case studies for Mozambique and Peru

161 Case study for Liberia

162 Case study for Cambodia

capped by the war) and encouraged him to go and find work, which led him to leave home and join the local militia. [Boy] 16, the child of his father's first marriage, constantly in conflict with his stepmother, left home to work in a bakery where his workmates, who were in the militia, encouraged him to join.[163]

"... when I was going [to the army base] there were two minors, one of 15 and one of 11 or 12. The 11 year old went to [the army base] because he had no family ..."[164]

... a 12 year old who was conscripted at ten years of age with his older brothers. This was done at the behest of his father who had remarried after the death of his first wife and no longer wanted the responsibility of raising his sons.[165]

[163] Case study for Lebanon (translated from the French)

[164] Testimony of a minor recruited but then freed by the army, quoted in the case study for Peru (translated from the Spanish)

[165] Case study for Cambodia

CHAPTER 4

Treatment as Soldiers

As children, they are suspects and probable victims. As soldiers, they are too often killed...[166]

Generally, child soldiers in government armed forces receive the same treatment and training as adult recruits:[167]

All my sources agreed that child soldiers received no special treatment in the armed units. There were no differences in the treatment of adult and child soldiers.[168]

According to the testimony of the minors, they were treated like any other soldier; in the military units in which they served no one took account of the fact that they were under age.[169]

Child soldiers were not accorded any preferential treatment to that of adult soldiers. Nor were there any variations in the type of activity they performed on the basis of their age or method of recruitment.[170]

In a great number of the interviews conducted ... teenagers reported that their duties were similar to those of all rank and file soldiers. [Government] soldiers in particular stated that distribution of duties

166 Case study for Turkey

167 Case studies for Afghanistan, Burma/Myanmar, Cambodia, El Salvador, Ethiopia, Guatemala, Nicaragua and the former Yugoslavia

168 Case study for the former Yugoslavia

169 Case study for Nicaragua (translated from the Spanish)

170 Case study for Ethiopia

had little to do with age. All ... we interviewed had been in active combat and had spent most of their tenure in the front lines doing the same fighting as the older soldiers. They had all fired weapons, and most had killed people.[171]

In those situations where the recruitment of children is illegal or unacknowledged it would be remarkable if this were not the case. In some respects equal treatment may be beneficial for the children in that they get paid and if their recruitment was in fact conscription, even though underage, they may be able to leave after the prescribed term of service, and receive the other benefits of normal service. (In armed opposition groups it is very rare for combatants, even children, to be allowed to leave unless invalided out).

However, the treatment of governmental recruits of all ages is often inhuman and degrading. This starts with initiation rites for new recruits ("baptism", "hazing") involving beatings, humiliation, being treated as servants, and being introduced to prostitutes and alcohol. The degree of violence and degradation involved not infrequently leads to death (including suicide), disablement or permanent physical, mental and emotional damage.[172] That such treatment is meted out in a state institution raises questions about human rights and governmental responsibility, as well as how such treatment affects the subsequent behaviour of soldiers. These are subjects which merit a study in their own right.

A few examples of such treatment suffice to illustrate the point:

... a succession of exercises lasting several hours to break the physical resistance of the recruit by provoking intolerable muscular suffering ... Hitting and body bruising with instruments such as machetes, sticks, rifle or pistol butts and yataganes (special clubs that leave no marks) on sensitive parts of the body; burning with cigarettes on various parts of the body or swallowing burning cigarettes; fist punches in the face, the nape of the neck, the ears; kicking in the legs and stomach and stamp-

[171] Case study for Burma/Myanmar

[172] Case studies for Burma/Myanmar, El Salvador, Guatemala, Honduras and Paraguay

ing on hands. Psychological pressures including threats, mockery and continuous insults ...[173]

... they are trained to become accustomed to sadism and, in order to lose their fear, they practise by cutting the throats of domestic animals and it is known that some of them are obliged to drink their blood. The internal violence which the recruits most fear is the verbal pressure exercised by the officers, pressure aimed at minimising their self-esteem and breaking their characters.[174]

Among the toughest exercises is "the ram" which consists of rolling on any surface which is stony, thorny, or asphalted etc. Your body ends up skinned because they have you just in your underpants. Also, while you are rolling, the man in charge of the squad kicks you or hits you with a club. Another difficult exercise is to be shut up in a waterproof tent, it consists of shutting up 20 or 40 recruits in the tent and then they throw in tear gas grenades and outside the corporals stand guard to make sure no one comes out until they give the order.[175]

Although such brutal treatment may be the normal experience of recruits, it may fall unequally on the child recruits who are less well-equipped physically, mentally and emotionally to withstand it:

... most child soldiers interviewed for this survey seemed to either fear or revere their commanders, and stated that they would follow their every order. Fear is a major determining factor in their obedience and performance. However, in extreme cases, when child soldiers could no longer tolerate their own, or others' mistreatment by their senior officers, they were driven to either suicide or murder.[176]

Or the child soldiers may be deliberately picked on:

[173] Case study for Paraguay (translated from the Spanish)
[174] Case study for Colombia (translated from the Spanish)
[175] Case study for Honduras (translated from the Spanish)
[176] Case study for Burma/Myanmar

"In most cases child soldiers received harsher treatment and were also blamed for faults of adult soldiers."[177]

"Irresponsible children" may be blamed by adult soldiers for acts of pillage carried out by the adult soldiers themselves.[178] The assignment of poor and inadequate food and medical care (as the lowest rungs on the military ladder) may be more serious for children, whose bodies are still growing and may be weakened by the exactions of military life:[179]

... the food is vile and also before and after eating we had to do knees-bends so that we would vomit our food ...[180]

The same recruit can do three shifts of guard duty in a single day, which means he can be standing up for nine hours, and exposed during the same period to guerilla attack.[181]

Treatment in armed opposition groups is equally inhuman in many instances, with those who cannot keep up or who try to escape being routinely killed so as not to reveal any secrets or to discourage others.[182] Even if such treatment is the same as for adults, "it is mostly children who suffer this fate as very few of them manage to stand up to the intensive training to which they are subjected, the military discipline, the diet and the unhealthiness of the surroundings."[183] Similarly, being less adept at looking after themselves, or standing up for their rights, they are more prone to die from starvation and preventable diseases contracted in the unhygienic conditions in which they live.[184]

However, this is not always the case:

"Even the most ardent critics of the insurgent [opposition] forces could not find fault with the manner in which child soldiers were treated."[185]

[177] Case study for Afghanistan
[178] Case study for Lebanon
[179] Case study for Burma/Myanmar
[180] Case study for Honduras (translated from the Spanish)
[181] Case study for Colombia (translated from the Spanish)
[182] Case studies for Burundi, Peru and Uganda
[183] Case study for Burundi (translated from the French)
[184] Case study for Ethiopia
[185] Case study for Chechnya

Where children are well treated in armed opposition groups, this is because all members of the group are well treated, reflecting "the warmth of friendship/camaraderie";[186] the need to maintain the numbers and morale of the group, the apparent absence of forced recruitment, and the ethnic- and community-based nature of the conflict. Similarly, where whole families are involved in the armed opposition group, the children tend to be treated well since they are under the care of their own families.[187]

DIFFERENTIAL TREATMENT

Some opposition groups are exceptional in seeking to take account of the particular needs and abilities of children. This may be by annually evaluating the child's physical and mental state when deciding on tasks and promotions, according children a degree of emotional attention, with some being "adopted" by commanders, providing time for play, and some preferential food and accommodation. Some, at least when there is a truce or a lull in the fighting, provide basic education.[188] Education, however, may have an ideological content:

> *... a third group was made up of the "legions" or "pioneers of the future", children under ten years old, of both sexes, who lived separated from their parents and attended the "people's schools" where they were given courses in political indoctrination.*[189]

Governments which legally recruit under-18s may also take special measures, for example, by having a policy of not sending them on active service[190] or "Recreational facilities are provided in an environment free from alcohol use. Their progress is carefully monitored by supervisory staff and there is ready access to chaplains, psychologists and other support staff."[191]

186 Case study for the Philippines
187 Case study for Afghanistan
188 Case studies for Colombia, El Salvador, the Philippines and Turkey
189 Case study for Peru (translated from the Spanish)
190 For example, Australia and the Netherlands
191 Response from Headquarters Australian Defence Force to enquiry by Quaker UN Office, Geneva

On the other hand, some armed opposition groups mete out especially cruel or dehumanising treatment to them. The use of children as executioners is not unusual,[192] although acts of ritual cannibalism appear to be the exclusive province of a small number of armed opposition groups seeking to imbue children with contempt for human life:

> *... the minors were obliged to cut the throats of those who had been declared culpable by the "people's courts". Often they were made to eat the entrails – heart, liver, kidneys – and to drink the blood of the rebels who were sentenced to death.*[193]

> *During [the boy's] time with the guerillas he had the opportunity of witnessing many executions of boys and girls of 12 and 13. Later, with the passage of time, he carried them out himself and taught others to do it. The practice of drinking the blood of the dead was a great trauma for him; he used to dream that the dead person would come back to reclaim his blood.*[194]

> *While, in the testimonies, there is mention of human targets to "toughen up" the children, it is important to add that in some cases these targets were parents of the child being trained. In other cases, the child could be forced to watch the ritual which might involve his/her parent. Such was the case of one child who stated: "In captivity, my father was used as a target during the final tests of boys who were being trained."*[195]

> *Almost all the children at one time or other were ordered to torture, maim or kill other children or adults who attempted to escape or to destroy property like burning up houses. With intimidation, threats and sometimes torture, children were obeying like robots.*[196]

[192] Case studies for Burma/Myanmar, Colombia, Honduras, Liberia, Mozambique, Peru and Uganda

[193] Case study for Peru (translated from the Spanish)

[194] Case study for Colombia (translated from the Spanish). Cannibalism being used as a rite of passage to combatant status by Renamo is reported in N. Boothby, P. Upton & A. Sultan, *Boy Soldiers of Mozambique*, Refugee Participation Network 12 (March 1992), pp 3–6, p 5

[195] Case study from Mozambique (translated from the Portuguese)

[196] Case study for Uganda

... reports indicate use of children as young as 10 years as assassins. Children of 12–14 years of age are also reported to have been used to massacre women and children in remote rural villages.[197]

There are numerous reports of children being given drugs and/or alcohol routinely or before battles.[198] Sexual abuse (particularly of girls) is not infrequently reported though by no means universal.[199] By contrast, some armed opposition groups ban the use of drugs and alcohol and demand respect for women.[200]

FUNCTIONS PERFORMED

In armed opposition groups children often start out in support rather than combat functions. They may, indeed, be "associated with the military" rather than an actual part of it, being alone and destitute and having attached themselves to individual soldiers or troops, and becoming servants or guards in exchange for food and shelter.[201] In many situations children are used for guard duty, patrolling and manning check-points:[202]

... most children ... were used normally for behind-the-front activities before being given military training and accepted as full-time combatants.[203]

As soon as they were recruited they would be given the duty of carrying luggage, for example looted foodstuffs including chickens, ammunition, property as well as injured rebels ... Most times the children did guard duties or worked in the gardens or were sent to hunt for wild fruits and

197 Case study for Sri Lanka

198 Case studies for Afghanistan, Burma/Myanmar, El Salvador, Honduras, Liberia, Mozambique and Peru; see also Amnesty International, "Sierra Leone: Human rights abuses in a war against civilians" (13 September 1995), p. 25

199 Case studies for Afghanistan and Mozambique; case study for Uganda in M. McCallin: *The reintegration of young ex-combatants into civilian life* (ILO, Geneva, 1995)

200 Case studies for El Salvador, the Philippines and Uganda

201 Case study for Rwanda

202 Case studies for Bhutan, Burma/Myanmar, Cambodia, El Salvador, Ethiopia, Guatemala, Liberia, Mozambique and Peru

203 Case study for Turkey

vegetables or looted food from gardens and granaries. But if attacked by the enemy, they were expected to fight back with the adult rebels. When there were fierce battles and shortages of fighters, child soldiers would be taken to the war front to fight alongside the adults.[204]

In any case, they were being prepared to play an active part when older:

Almost all the children who stayed for over one month [before escape or capture] were given military training of parade, taking cover, assembling and dismantling a gun, shooting, and offensive tactics.[205]

Children from 10 years of age could become combatants, although mainly children from 14 or 15 years up participated in fighting with arms.[206]

... around 1,000 children aged between 7 and 14 held a meeting during which a decision was jointly taken to begin military organising activities and establish childrens' committees which will be similar to the guerilla organisation. The decision ruled that these committees will organise for the time being on the training level and that, after the age of 14, volunteers would transfer to the ranks of the guerillas.[207]

Children have also been observed to spend one or two hours a day out of school digging bunkers as a form of militarised civic duty and then motivated to enlist and join the armed opposition group.[208]

Numerous children are used as porters during conflicts. Often they carry very heavy loads (food, up to 60 kg of ammunition, injured rebels) and receive severe treatment at the hands of their superiors. In some cases, they are savagely beaten and, if they become too weak to carry their load, liable to be shot.[209]

Many of the case studies refer to a special preference for using children

[204] Case study for Uganda
[205] Case study for Uganda
[206] Case study for El Salvador
[207] Case study for Turkey
[208] Case study for Sri Lanka
[209] Case studies for Burma/Myanmar, El Salvador, Mozambique, Sri Lanka, Turkey and Uganda

as look outs, messengers and for intelligence work:[210]

There is also evidence that very young soldiers or civilians are used as spies in the disputed areas. They are generally given cows to herd and asked to wander around and to report back on any [opposition] activity they had witnessed. In [one area], for example, children herding cows were asked by the [opposition] to place home-made landmines (crude bombs made from fertilizer) on rural roads late at night or early in the morning. In some cases, the local government forces will use young civilians or very young soldiers to do the same near [opposition] villages. Very young soldiers are not noticed so much and can easily pass themselves off as normal peasant children and are used by both sides as intelligence gatherers. ... Children were used a lot to lay and to clear mines but we did not hear of them being used exclusively for this purpose.[211]

... they are likely to be less noticeable than adults. Because of their size, they can hide more easily, and they have more innocent appearances ... [An NGO] interviewed a 14-year-old [opposition] recruit in 1990, who told an autobiographical story that seemed normal. The next day he was caught in the ammunition supply room ... it was discovered that the [government forces] recruited him as a spy and were threatening him that they would kill his grandmother who had taken care of him since he was a baby. He was later executed by the opposition group concerned for spying, despite his young age.[212]

[In the opposition forces] younger children were considered more suitable as messengers, since they would not raise suspicion ... In the National Guard, the National Police and the Treasury Police, minors

210 Case studies for Afghanistan, Burma/Myanmar, Burundi, Cambodia, Colombia, El Salvador, Ethiopia, Guatemala, Liberia, Mozambique, Nicaragua, Northern Ireland, Peru, the Philippines, South Africa, Sri Lanka and Uganda; see also case study for Uganda in M. McCallin: *The reintegration of young ex-combatants into civilian life* (ILO, Geneva, 1995)

211 Case study for Cambodia

212 Case study for Burma/Myanmar

generally did not bear arms but performed other tasks, for instance, serving as "ears", informers, messengers etc.[213]

... children were used by the [opposition] force to run espionage and infiltration missions. Children of tender age were used to carry out these missions.[214]

... the smaller male child soldiers are used to carry out household chores, information gathering/intelligence, running of errands and setting up of ambushes. In this case, the child soldiers are posted on highways and strategic places under the pretext that they are desperately in need of help/assistance. When the opposing forces come in contact with them, they in turn signal to their group for attack and in the process they (child soldiers) are sometimes killed.[215]

Not surprisingly, where the recruitment of children is prevalent, including their employment as spies, this tends to have a detrimental effect on the safety of all children in the conflict area:

[Government] soldiers armed with knives, machine guns, rifles and grenades assassinated 69 peasants, among them 21 children under 5 and another ten between five and ten. An interview is quoted with Sub-Lieutenant ... in which he affirms that even the children were "dangerous", because ... "guerillas begin to indocrinate them at two, three or four years [age], getting them to carry things and taking them to different places."[216]

The deliberate killing of children in this way, as well as the repeated references to children being killed by government armed forces when trying to evade recruitment, during training and military service and on capture, raises serious questions about government compliance with the requirements of international human rights law. (Leaving aside the questions of

[213] Case study for El Salvador
[214] Case study for Ethiopia
[215] Case study for Liberia
[216] Amnesty International Report quoted in case study for Peru (translated from the Spanish)

international humanitarian law). Even if they are alleged to have committed an offence, the age of criminal responsibility needs to be taken into account, and they are entitled to the rights of due process. In any case those under 18 years of age at the time of the commission of the offence should not be executed.[217]

Size and agility (and greater expendability) also mean that children are given particularly hazardous assignments:

> *"I was in the front lines the whole time I was with the [opposition force]. I used to be assigned to plant mines in areas the enemy passed through. They used us for reconnaissance and other things like that because if you're a child the enemy doesn't notice you much; nor do the villagers."*
>
> *"... when I was sixteen. My job was to run out into the "Killing Fields" [no man's land] and grab weapons, watches, wallets and any ammunition from the dead soldiers, and bring it back to the bunkers ... This was a difficult job as you could see the enemy and they could easily "pick you off" as you ran out and back again. The major and I had to do this job as we were the smallest."*[218]
>
> *Children are used to detect explosive weapons (bombs, claymore mines, etc) ...*[219]
>
> *Young soldiers normally stayed in front of the others and they were normally used to demine, or to observe the situation.*[220]

Suicide missions are considered to be the particular province of adolescents:

> *They are more physically suited for carrying out such operations and are better disposed to it mentally.*[221]

[217] International Covenant on Civil and Political Rights, Article 6(5), a non-derogable right; Convention on the Rights of the Child, Article 37(a)

[218] Both testimonies of opposition child soldiers quoted in the case study for Burma/Myanmar

[219] Case study for Guatemala (translated from the Spanish)

[220] Case study for Cambodia

[221] Case study for Lebanon (translated from the French)

Urban volunteers act as "instigators, agitators, defenders of the revolutionary principles" as well as participating in acts of sabotage.[222]

However, children are used in the front lines either routinely or if required for a large-scale battle.[223] Inexperience and lack of training results in a high number of casualties: "these children are massacred like flies".[224]

One commander observed:

> *"When there is shelling, the younger ones forget to take cover. They get too excited. They have to be ordered to get down inside the bunkers."*[225]

A journalist, having witnessed the death of an 11-year-old recruit who was "virtually cut in half by a 120 mm mortar explosion" noted:

> *"He did not duck. He didn't seem to understand the ferocity of these weapons."*[226]

There are reports of large numbers of child soldiers, often plied with alcohol and drugs, being used in human wave attacks:

> *... there were a lot of boys rushing in the field, screaming like banshees when they rushed the barbed wire (barricade in the field). It seemed at first like they were immortal or impervious or something, because we shot at them but they just kept coming.*[227]

> *... in a major battle ... children were used for a massed frontal attack.*[228]

A number of child soldiers are forced to commit atrocities against their own family or community in order to cut off the possibility of return,[229] or

[222] Case study for Guatemala

[223] Case studies for Afghanistan, Burma/Myanmar, Burundi, Cambodia, Chechnya, Colombia, El Salvador, Ethiopia, Guatemala, Liberia, Mozambique, Turkey and Uganda

[224] Case study for Burundi (translated from the French)

[225] Case study for Burma/Myanmar

[226] Case study for Burma/Myanmar

[227] Case study for Burma/Myanmar

[228] Case study for Sri Lanka

[229] Case studies for Afghanistan, Bhutan, Burma/Myanmar, Colombia, Guatemala, Honduras, Mozambique and Nicaragua

to participate in operations against the peasant community, where many were born and raised:

There were some who took part in clean-up operations carried out by [the government forces] in the poor districts of the country's cities and in the rural zones. These consisted in searching houses suspected of sheltering [opposition] guerillas, arrests, beatings, imprisonments, tortures and, in not a few cases, the disappearance of young people suspected of carrying out subversive activities.[230]

One report cites the use of under-age soldiers to undertake labour, under the same disciplinary military regime and without payment, in the homes and businesses of military officers.[231]

In urban situations, children are involved in street resistance, riots, throwing petrol bombs or stones at law enforcement personnel, and so on.[232]

A particular task assigned to children may be the recruitment of other children, whether by persuasion, threats or kidnapping by force.[233]

PUNISHMENT

It is not always possible to distinguish routine treatment from specific punishment. There are frequent reports of beatings by superiors, simply as random occurrences but sometimes as punishment:

Sometimes when I fell asleep when I was on sentry duty, I was beaten by my corporal. He beat me like a dog, like I was an animal, not a human being. There were two or three suicides during that time; of boys who had been hospitalised and finally shot themselves.[234]

Some discipline the children by means of verbal reprimands or the assignment of practical tasks, such as digging holes for rubbish, cutting wood and

230 Case study for Nicaragua (translated from the Spanish)

231 Case study for Paraguay

232 Case studies for the Intifada, Northern Ireland and South Africa

233 Case study for Peru

234 Case study for Burma/Myanmar

giving courses in politics to new recruits. Others dole out extra physical training with the intensity varying in accordance with the gravity of the offence, for example, push-ups for something less serious, two hours or many days of forced exercise for a more serious offence:[235]

> *... one 15-year-old soldier ... was playful and often irresponsible. Due to his negligence during duty hours he was regularly punished (front/back-roll on the ground or carrying sandbags for several hours).*[236]

Many child soldiers receive severe physical abuse for failing at military training, falling asleep on sentry duty or disobedience.[237] In one situation child soldiers were reported to receive 150 lashes on the back, neck, head and legs for "lesser crimes".[238]

> *As soldiers in the [opposition] bases, the children were severely punished if they disobeyed the orders of their superiors. The punishments varied from simple corporal punishment, deprivation of food, amputation of fingers, nose, ears, and even execution carried out by one of the more "mature" children.*[239]

> *... the "Tripod" which consists of supporting the body on the tips of the toes, with the hands behind and the head resting on some uneven surface (gravel, bottle tops, a paving stone, etc) for as long as the superiors order.*[240]

In government armed forces, military law applies to all soldiers. The implications of this for child soldiers deserve further consideration.

For attempted escape and disobedience children are beaten to death or shot. Many case studies report children being shot for trying to escape from recruitment or afterwards:[241]

[235] Case studies for Colombia, El Salvador, Paraguay and the Philippines
[236] Case study for Bhutan
[237] Case studies for Burma/Myanmar, Honduras and Paraguay
[238] Case study for Uganda
[239] Case study for Mozambique (translated from the Portuguese)
[240] Case study for Honduras (translated from the Spanish)
[241] Case studies for Colombia, Ethiopia, Liberia and Uganda

Soldiers were regularly shot, accused of wounding themselves in order to be hospitalised, retreating in the face of the enemy, or simply gossiping.[242]

Generally, punishment is to be sent to the front line or to have salary withheld but discipline in the army is so poor that any attempt to impose sanctions or punishment is met by desertion or by violence. Reports in the [press] during the 1994 offensive ... featured chilling accounts of out-of-control troops killing their own officers.[243]

In at least one opposition group, members are shot for stealing.[244] Even where death is not inflicted directly it figures frequently as the indirect consequence of other punishments. The repeated references to the killing of children at all stages in the process raise the question of how many of those tapped for recruitment ever make it to the front line, let alone survive to be demobilised.

DEATHS, INJURIES AND HEALTH CARE

... young soldiers are more likely to get killed or injured than adults because they are braver, they tend not to know their locations as well as adults, have often been less well trained or simply have less survival experience.[245]

In most cases, children receive the same training and health care as other soldiers. What this means depends on the resources available and the state of the conflict at any given time. New recruits who are injured, fatigued or simply hungry are often shot and left to rot.[246] Sometimes, both government and opposition groups, during extremely heavy fighting, simply leave the wounded on the battlefield.[247] At other times, badly wounded soldiers are

[242] Case study for Ethiopia
[243] Case study for Cambodia
[244] Case study for Colombia
[245] Case study for Cambodia
[246] Case studies for Burundi and Uganda
[247] Case study for Guatemala

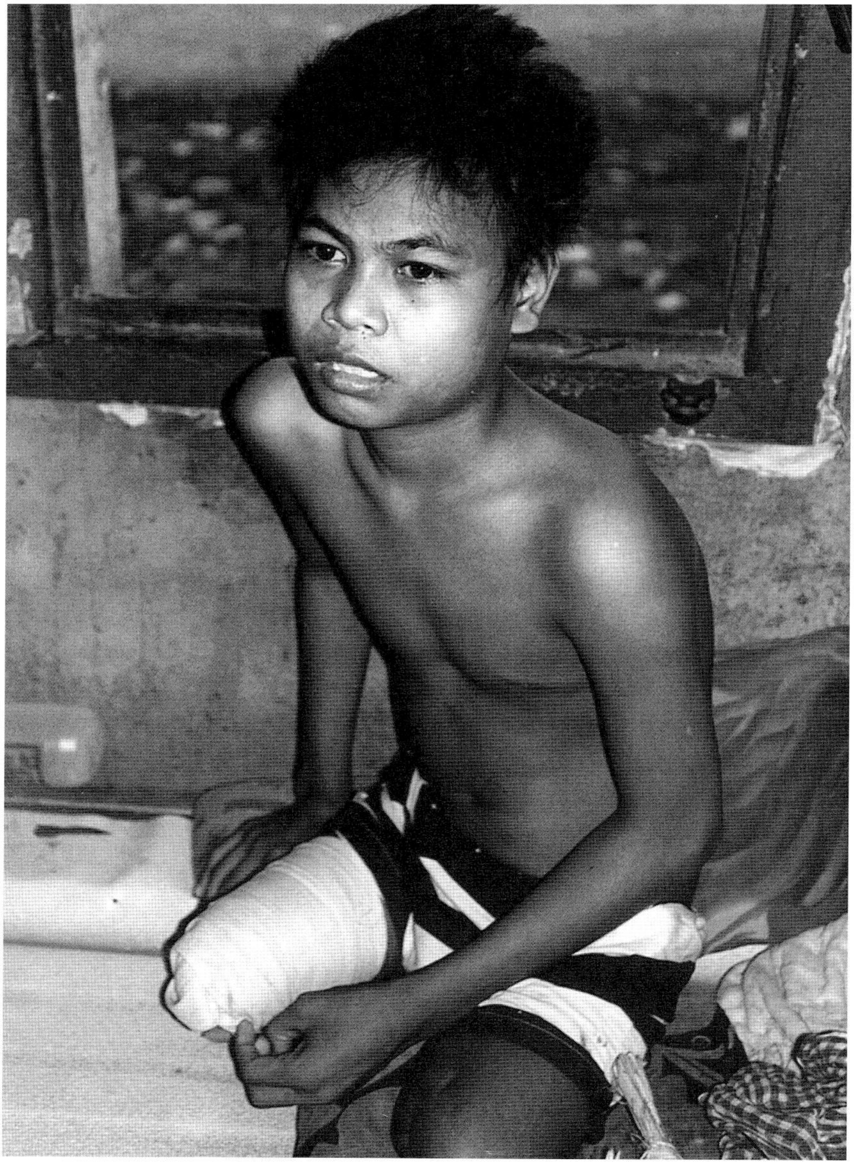

This Cambodian boy joined the army at the age of 15. He fought the Khmer Rouge for four years, until he stepped on a landmine in 1995 and lost both legs.
Photo: Carl von Essen.

"just killed".[248] On the other hand, they may be treated by doctors,[249] at military hospitals, or in barracks by the "medical team".[250] Sometimes, armed opposition groups send injured children home to their families or to foster families to be looked after, or they are taken to the outskirts of settlements so that they can be taken in by the local community.[251] At times, the only medicine available is herbal.[252]

> *Through a lack of medical services up to the task of meeting the vast needs, child soldiers who were injured, mainly through mines, and shrapnel from other explosives, ended up losing limbs which were amputated, thus not receiving the necessary medical care.*
>
> *... many [demobilised child soldiers] complained of health problems related to the presence of foreign bodies in their organs (bullets, shrapnel ...). This is a problem as many families do not have the resources to pay for operations to remove these objects.*[253]

There are also reports of medical supplies being reserved for officers or stolen by superiors, either to be sold and replaced by cheaper supplies or for personal profit.[254] In other cases, financial and other care is provided for some who are seriously injured in battle, as well as for families of war victims.[255]

The most frequent references to specific injuries suffered by child soldiers are to loss of hearing, loss of limbs and blindness:[256]

> *The main injuries received by the children are deafness, blindness, burns, damaged limbs leading to amputation, given the frail nature of*

[248] Case studies for Burma/Myanmar and Cambodia
[249] Case studies for Colombia and Sri Lanka
[250] Case studies for Cambodia, El Salvador, Ethiopia, Guatemala, the Philippines, Turkey and Uganda. The qualifications of the medics vary significantly.
[251] Case studies for El Salvador, Guatemala, Mozambique and the Philippines
[252] Case studies for Mozambique and Uganda
[253] Both from the case study for Mozambique (translated from the Portuguese)
[254] Case study for Burma/Myanmar
[255] Case studies for Guatemala and Turkey
[256] Case studies for Afghanistan, Burma/Myanmar, Cambodia, El Salvador, Guatemala, Lebanon, Mozambique and Uganda

their bodies, and from the hazards of carrying heavy weights, inhaling toxic substances, land mines, and long marches, to name but a few.[257]

In a direct interview with an ex-soldier, the principal causes of death and injury of minors enrolled in the army were stated to be the explosion of claymore, anti-personnel and anti-tank mines placed by the guerillas. This situation is due, according to the ex-soldier, to the use of minors as advance scouts and as mine detectors. Other causes of death and injury among minors were the explosion of grenades, rockets and bombs. These explosions cause serious injuries and permanent damage, such as the mutilation of limbs, deafness, blindness, total or partial paralysis and on other occasions death.[258]

[A 16 year old] was involved in all the battles and very soon lost his hearing due to his having fired a B7 cannon (105 mm without recoil) non-stop during a battle.[259]

Mine injuries remain among the most common. The other injuries are from the shrapnel of explosions from shells, mortars, B-40s as well as bullet wounds. Losing a foot or a whole leg or both legs from landmine explosions is common. The loss of arms and hands, lacerations to the body, loss of eyesight and hearing are also common injuries among soldiers.[260]

There are conflicting claims about which children are most at risk of death or injury:

Such injuries were most prevalent in the inexperienced child soldiers, and, therefore, the most affected are boys between 11 and 15.

Some suggest that it is the youngest and most inexperienced. Others suggest that these are kept behind the lines and therefore those in the next

257 Case study for Mozambique (translated from the Portuguese)
258 Case study for Guatemala (translated from the Spanish)
259 Case study for Lebanon (translated from the French)
260 Case study for Cambodia

age group are most vulnerable. In all cases, the children were considered to be at greater risk than adult and experienced soldiers:

> *Carelessness, insufficient training and over-excitation were the principal causes of the increased death toll amongst child soldiers.*[261]

> *There were a lot of injuries due to: (a) fighting or quarrelling between themselves, (b) battlefield injuries and (c) drowning when crossing rivers while on patrol (the young, small boys could not wade in the deep water like the adults while carrying heavy materials like packs or guns). In his battalion, of 30 child soldiers, 15 died or were killed.*[262]

In any event, age is relative and all those falling into the above categories are still children whose bodies have not fully developed. Consideration therefore needs to be given to the additional emotional, psychological, physical, educational, economic and social disadvantages entailed as compared to adults. Loss of sight or hearing may be severe impediments to future educational, vocational or social development. Loss of limbs may sometimes entail repeated amputations for those still growing since the bone of the amputation site grows more than the surrounding tissue, and they will also require new prostheses frequently.[263] In addition to the trauma involved, costs may be too high or the necessary facilities may be unavailable. In societies with high levels of unemployment, the additional disadvantages of such handicaps may be the last straw (see Chapter 5).

TREATMENT ON CAPTURE

> *It so happens that in many instances, under-age suspects never do reach [the] courts and are often nothing more than the usual daily news headlines on [government] television: "Troops have killed so many terrorists today."*[264]

[261] Case study for Afghanistan
[262] Case study for Cambodia
[263] S. Roberts and J. Williams: *After the Guns Fall Silent* (Vietnam Veterans of America Foundation, Washington DC, 1995), p 10
[264] Case study for Turkey

Governmental treatment of captured child soldiers varies enormously. Many are treated the same as captured adult soldiers which, in the context of the situations considered for this report, means that they may be treated as criminals or terrorists[265] or they may be held in military prisons.[266] Many captured child soldiers (of both sexes) are subjected to abusive interrogation procedures, torture, taunts, isolation, rape with various implements, being bound in the "banana" position, promises of rehabilitation or threats of death, and being held for prolonged periods of detention or imprisonment. Some die as a result.[267]

Some government armies and armed opposition groups retrain captured child soldiers and incorporate them into their own ranks or give this as an option.[268] This is a distinctive feature of the lives of child soldiers as opposed to adult soldiers and presumably reflects the greater acceptance of participation as normal or inevitable by the children themselves and of the commanders' belief that the children can be retrained to participate on the opposite side.

The fate of those caught spying, however young, is usually death, sometimes accompanied with torture, even if other captured child soldiers are treated as prisoners of war "and render duties such as digging trenches and logistic services".[269]

The inadequacy of detention facilities and lack of adherence to the concept of due process affect children and adults alike but, in particular, there are rarely special facilities and protection for children in such circumstances,[270] although this is required by the Convention on the Rights of the Child, the International Covenant on Civil and Political Rights and other

[265] Case studies for Lebanon, the Philippines, Rwanda and Turkey

[266] Case study for Lebanon

[267] Case studies for El Salvador, Ethiopia, the Intifada, Guatemala and Turkey

[268] Case studies for Afghanistan, Cambodia, Ethiopia and Mozambique

[269] Case study for Afghanistan

[270] Case studies for the Intifada, Lebanon, the Philippines, Rwanda and Turkey; see also Amnesty International, "Sierra Leone: Prisoners of war? Children detained in barracks and prison" (12 August 1993)

instruments.[271] Long delays in the administration of justice impact severely on children because of the deprivation of family or substitute care, education, poor food and health provision, and so on. This is one of the reasons why Article 37 of the Convention on the Rights of the Child provides: "The arrest, detention or imprisonment of a child shall be used only as a measure of last resort and for the shortest appropriate period of time." Social welfare institutions may be reluctant to take them on fearing that they will be a bad influence on other children being accommodated.[272]

Detention and questioning of children also occurs when there is no specific evidence of their participation in the armed conflict:

> *... the army's counter-insurgency operations involve the systematic detention of men, women and children found in areas recaptured from rebels. Young people, children and babies have been caught up in the army's efforts to identify rebels and those who may have helped or collaborated with them.*[273]

In addition to the random detention of children as part of the civilian population in this way, where children have been recruited into armed opposition groups, those in the relevant age group may be detained:

> *Often, the only grounds for detention appeared to be a general suspicion that men and boys of fighting age may have fought with the rebels.*[274]

This illustrates the problem that, as long as children are recruited into armed forces or armed opposition groups, they themselves and other children cannot be insulated from the consequences of, or the suspicion of, involvement.

Age was a consideration in the minds of some captors. Some have a pol-

[271] Convention on the Rights of the Child Articles 37 and 40, International Covenant on Civil and Political Rights Articles 9, 10, 14, and 6 (in relation to the death penalty), and the UN Standard Minimum Rules for the Administration of Juvenile Justice (known as "The Beijing Rules")

[272] Case study for the Philippines

[273] Amnesty International: "Sierra Leone: Prisoners of war?" (12 August 1993)

[274] ibid

icy of treating child soldiers as innocent victims: trying to scare off adult soldiers in order to gather up the children without having to fight them, questioning them for intelligence and to ascertain information necessary for family reunification, and then handing them over to others for processing, although the policy may not always be implemented in practice.[275] One interviewee noted that some minors captured by the government forces were not killed and concluded that this may have been because of their age as no adult soldiers were "saved".[276] There is always a danger that governments will use captured child soldiers for propaganda purposes.[277]

It may only be when captured and charged that the real age of the soldier becomes known, either because this is the first time proof of age is sought[278] and/or because there is an incentive for the accused to reveal it when faced with criminal proceedings:

> *Many instances arose where youths (under 17) lied about their age to get into the [armed opposition group] – in lots of cases their true age only became known when they were charged.*[279]

> *... a soldier who accidentally killed his army friend and who, when brought before a military court, did not hesitate to present his real age so that his case would be transferred to the juvenile court where punishments are automatically reduced.*[280]

One of the consequences of child recruitment into armed opposition groups can be that the age of criminal responsibility is reduced and/or that anti-terrorist or emergency legislation is introduced which takes no account of the age of suspects, or reduces the age of applicability below that for "normal" criminal offences:[281]

[275] Case studies for Mozambique and Uganda
[276] Case study for El Salvador
[277] Case study for Turkey
[278] Case study for Guatemala
[279] Case study for Northern Ireland
[280] Case study for Lebanon (translated from the French)
[281] Case studies for the Intifada, Nicaragua, Peru and Turkey

Children were regularly interrogated using the same physical and psychological pressures that are applied to adults.[282]

Although this may be seen as a normal reaction for a government faced with this particular problem, it raises questions about compliance with the provisions of the Convention on the Rights of the Child, and of the adequacy of protection and safeguards for child suspects, such as access to a lawyer, suitable rehabilitative provision for those found guilty and the non-application of the death penalty to those under the age of 18 at the time of the commission of the offence. The question of the responsibility of child soldiers for their actions, the age of criminal responsibility, appropriate determination procedures and disposal arrangements need further consideration.

[282] Case study for the Intifada

CHAPTER 5

Demobilisation, Rehabilitation and Social Reintegration

THE PURPOSE OF THE Child Soldiers Research Project was to develop a better understanding of both the causes and consequences of children's participation in armed conflict. It is at the time of demobilisation, and the children's return to their families and communities, that the profound effects of their participation become apparent. The information from the research indicates that the children will confront a situation of multiple risks which can significantly compromise their reintegration into civil society. This situation is a function of the events they experience as soldiers, as described above, but is also due to the fact that the living conditions of their families and communities, which in most cases occasioned their participation in the first place, are even more precarious than when the children were recruited.

In outlining various issues that were raised in the case studies, this chapter discusses the consequences for the children as a continuum that includes their experience before and after recruitment, not only their experiences as soldiers. We believe this is important, firstly to emphasise from the perspective of the children themselves how their experiences can affect their developmental processes, and thereby constitute a gross infringement of their rights within the terms of the UN Convention on the Rights of the Child (CRC).

In addition, it is only through an appreciation of the conditions of life which define the children's social and cultural environment upon demobilisation that effective strategies can be implemented to address the consequences of their participation, and effect their social reintegration. Such a framework incorporates attention to the reasons for the recruitment of children, as much as the consequences, through action which enables families and communities to protect their children. This is a theme that was emphasised in the case studies for the research, where repeated reference was made to the linkage between education, employment opportunities, and the economic security of the children's families as the factors that would not only determine successful social reintegration, but also contribute to efforts to prevent further recruitment.

DEMOBILISATION

The first stage in this process is obviously to secure the demobilisation of the children, and to ensure that this occurs in such a manner that programmes can be implemented to address their special needs. The information on the demobilisation of children provided for the research project in fact describes only limited instances where child soldiers have been demobilised, and where some effort was made to assist in their rehabilitation and reintegration.

The only example of an official demobilisation of children was described in the case study for Liberia. A small number of children (184, or 6% of the total number of combatants disarmed at that time) were included in a process of planned disarmament and demobilisation of combatants in March and April of 1994. The climate for demobilisation resulted from the signing of the Cotonou Agreement in July 1993, though this did not specifically include recognition of the role played by children in the conflict, nor specify strategies to assist in their rehabilitation and social reintegration. The emergence of factions who had not been signatories to this agreement, and the subsequent resumption of hostilities "complicated the already fragile demobilisation... and brought the process to a complete halt."[283]

Starting in March 1994 UNOMIL,[284] through its spontaneous demobil-

[283] Case study for Liberia

[284] United Nations Mission in Liberia

isation programme, demobilised a further 434 child soldiers aged 17 years and under. It would appear that almost all of these children were demobilised with an adult relative who had been serving with the same forces. This characteristic of service in Liberia has resulted in children returning directly to their communities of origin, with only a small number expressing a wish to take advantage of the programmes of rehabilitation that are available.[285] Subsequent demobilisation of children in any planned manner was considered to be dependent on the implementation of procedures for general disarmament and demobilisation negotiated through "the interim government ... local and international relief organisations, civil/interest groups, and church-related organisations."[286] [287]

Although not included as a case study for the research project, an official demobilisation of children during an on-going conflict occurred in Sierra Leone, which was followed by a programme to address their rehabilitation and social reintegration. In June 1993, in recognition of their obligations following ratification of the CRC, the government army demobilised 370 children. It should be noted that this also resulted from advocacy on the part of the local UNICEF office to release the children.

The case study for the Philippines describes demobilisation of children during conflict when they become "bored or demoralized" or when there

[285] There are also reports that in Liberia whole families move with certain factions. A similar involvement of families is described with the Mujahideen in Afghanistan. Whether such involvement reduces the effects of participation on children, and/or contributes to effective rehabilitation and social reintegration, is a matter for further enquiry. Some children with the PKK may also be with their families, as "most had been flushed out of their homes along with their families and had accompanied them to the camps where they joined the *freedom movement.*" (Case study for Turkey)

[286] Case study for Liberia

[287] In April 1996 fighting again broke out in Monrovia. At the time of finalising this book (June 1996) there was no peace initiative and, rather than progress in the demobilisation of children, it was considered more likely that there would be further recruitment of under-age combatants, including re-recruitment of children who had already been demobilised.

has been a "violation of disciplinary regulations."[288] This may be due to the "special consideration" given to these children who were serving with the NPA, where there is monitoring of their well-being. A contrast to this example is given in the case study for Sri Lanka, where demobilisation of children with the LTTE is not only impossible, but desertion also is prevented and punished.

> *Once recruited, a child cannot leave the LTTE. Their hair is cropped as soon as they are enlisted. Some deserters hide till their hair has grown so they cannot be discovered. Others who wanted to go back to their parents are given a sound thrashing before their parents and sometimes by their peers.*[289]

In Afghanistan "there is no planned demobilisation scheme to assist and facilitate the return of child soldiers to civilian life." In this instance, demobilisation of children was and is a function of the particular group with whom they were serving. Children who served in the forces of the pro-communist regime were required to serve for four years and were then demobilised, although no mention was made of any financial or other material benefits awarded at this time. They could also be released from service as a result of a disability – presumably as the result of injury during conflict. Desertion of children was a common phenomenon, and was prevented by sending them "to unfamiliar, remote areas far away from their homes." Children who fought with the Mujahideen between 1978 and 1992 could, in principle, leave whenever they wished. There was, however, encouragement for them to remain both from religious persons and their families. As in this instance the children's families were present with them, it would have been difficult for them to resist the social pressure that would seem to have been present. There is no specified period of service for children presently fighting with different warring factions (1992 to the present). The

[288] Also in the Philippines' case study there is reference to a rebel returnee programme for those who apply for the governmental amnesty, but there is no record in the case study of anyone receiving assistance under the terms of this programme.

[289] Case study for Sri Lanka

case writers in this instance consider that these children "have no option left except to stay in the army" as there is a lack of facilities, the children are poorly educated, and the army "at least provides them with earnings."

A disturbing form of demobilisation, and certainly not one to be advocated, is described in the case study for Colombia where, if a young recruit "manages to kill a 'subversive', he is demobilised and returned to civilian life, 'as a form of payment'."

In Burma/Myanmar, the case writers had "never encountered evidence of an official demobilisation procedure for child soldiers working with the Tatmadaw" and, despite "signings of a number of cease-fires, there has been no significant demobilisation of either government or opposition troops." As "troop numbers are made up mainly of very young people", the process of demobilisation here would seem to require not only recognition of the participation of children, but special considerations for rehabilitation and reintegration. It is not considered that this will happen as there is a denial that children are involved in the military.

In on-going conflicts there is little hope of demobilisation, even when this may have been specified in peace agreements, such as in the case of Guatemala. Here army conscripts are demobilised at the end of their term of service, if they have infringed military law,[290] or if they are disabled. The case study for Guatemala did also mention that under-age conscripts can be demobilised if they have proof of their age, but the numbers were not specified. The problem here may be that many children from impoverished or marginalised communities will not be in possession of the appropriate documents. No form of demobilisation is available for those serving with the guerilla movement. Their service ends with capture by the army, or

[290] Such an infringement may lead to prosecution and imprisonment, and thus also has implications for the legal process to be followed where the offender is under age. A related concern is noted in the case study for Peru where "adolescents supposedly responsible for the crime of terrorism could be condemned to various sentences, including life imprisonment." A fuller investigation of how child combatants are treated within the legal system is required, particularly the extent to which the system may be used to exact 'retribution' on the part of one faction.

through desertion or disability. A similar situation pertains in Turkey for children serving with the PKK. Their term of service is not specified and they remain until "they are killed or assigned at a later stage to peacetime activities (i.e. activities in Europe, publishing, propaganda etc.)."[291] The latter could be viewed as a form of demobilisation, but is obviously based on some form of selective process, and raises the question of *who* is considered suitable for these activities.

Certain of the case studies described forced conscription of under-age recruits to government forces where there is a set term of military service. Whether such under-age conscripts are in fact released at the end of their term of service is a question that needs further exploration. One case study stated that they do in fact "remain in military service longer than the time prescribed by the law (2 years), and have to claim or demand their release, otherwise no one remembers that the period has been served."[292] In El Salvador and Afghanistan, although the period of military service may have been extended during the conflict – in the latter case from two to four years under the pro-communist regime – release did take place upon completion of service.

The demobilisation of children during conflict is influenced by the reasons for their recruitment, whatever the manner in which this is conducted. Advocating for their demobilisation will in some cases confront complex social, political and cultural issues that are directly associated with the causes of the conflict, and/or the manner in which it is prosecuted. Where people "cannot escape" from a conflict that "is taking place in their communities and in their own homes",[293] they are unable to protect their children from recruitment and even less likely to be able to effect their demobilisation. This situation may be further compounded by cultural conceptions of appropriate roles and duties of children, particularly when specific groups or communities are targeted, and even how "childhood" is itself determined. The case study for Chechnya describes such an example,

[291] Case study for Turkey
[292] Case study for Honduras
[293] Case study for the Philippines

where "youths between the ages of 16 and 18 were not considered children" and thus their participation is considered "normal" and "natural". Participation of some children in the Lebanon is sanctioned by religious precepts "in accordance with which Jihad becomes an obligation for the religious faithful including children if they have reached the age of majority according to the chari'a, i.e. if they have reached puberty."[294]

The information from the case study for Afghanistan provides very clear examples of how demobilisation is a function of the form of recruitment and of which faction is responsible. In some cases the supportive attitude of their families and communities towards their participation[295] may prolong their involvement. Ideological commitment can be a significant factor in convincing people of the legitimacy of involving children, and even encouraging their participation. This is a particularly difficult context within which to introduce concepts of child rights and welfare as the basis for advocating for prevention of recruitment and demobilisation. It was an issue noted in certain case studies, but not discussed in any detail, and is certainly one that requires further investigation.

> *... a spirit of solidarity developed amongst the people to voluntarily join the Mujahideen (resistance) forces and offer the services of their young children. This was done mainly based on their religious beliefs.*[296]

> *For the MNLF, child soldiers are a natural outcome of their holy war ... Muslim male children are reared side by side with a gun. At nine they can dismantle and put back an armalite. (A TV) news feature on the MNLF shows families socializing children into the holy war from the time they are born. Especially if fathers or family members have been killed in the conflict. For them, the holy war is a noble commitment and a natural consequence of their religious beliefs and cultures.*[297]

Tacit approval, if not direct encouragement, for the involvement of children in conflict may also characterise the child's social environment. This will be

[294] Case study for the Lebanon
[295] Palestine; Philippines; Chechnya; El Salvador; Colombia
[296] Case study for Afghanistan
[297] Case study for the Philippines

the case particularly where there is glorification of the military as, for example, in Burma/Myanmar.

> *Children, especially young boys, are raised to revere military leaders of the past, and to look on military induction as a sign of manhood. In much of the popular media, the soldier is held up as the perfect role model ... To be a soldier is to occupy a position of great honour and self-sacrifice. The emotional pull of such prestige should not be underestimated.*[298]

It will be apparent that in such situations advocating for the demobilisation of children may be seen as a direct challenge to cultural beliefs.

People's living conditions may also be a significant factor that not only influences recruitment, but prevents the demobilisation of the children. From the perspective of those involved, such situations may result from haphazard conditions, as the conflict moves from one location to another, or from the ways in which the conflict affects the capacity of different groups to protect their children.

> *The means of protection that families have developed to keep their children out of the war ... depend on the different phases of the war and on the situation that a family found itself in: what was applicable to Lebanese was not applicable to families from ethnic or religious minorities or to refugees due to the fact that they do not have roots in the Lebanon and therefore no extended family or home village. (As the war progressed) impoverished families resigned to their fate saw the engagement of their sons in the militia as a means of protection, a last resort in the absence of any alternatives."*[299]

> *The need for self-defence as the reason for involvement in armed forces characterises those cases of child soldiers who lived near the front line and whose lives and the lives of their families were in danger because of the conflict. Sometimes these boys were the only male members of the*

[298] Case study for Burma/Myanmar
[299] Case study for Lebanon

family as the adult men were serving as soldiers in some other armed groups. This was mostly the situation in rural settings.[300]

Families may even see material advantage, regardless of any ideological commitment, in their children's involvement, and be reluctant to forego the benefits that the child combatants obtain for their families.

> *For many families allowing their children or volunteering to send their children to fight for the country was a good thing. According to* CAW *field experience many mothers have remarked on the joy of seeing their ten-year-old dressed in a brand new military attire carrying an AK-47. For some families the looted property that child soldiers brought home further convinced them of the need to send more children to the war front to augment scarce income.*[301]

Even in situations where communities *in principle* may not approve of children fighting, in some way these attitudes and beliefs are set aside in time of war, when they do "not have time to give much thought to these things".[302] Additionally, some children may join armed groups for protection because they have "lost both parents in the war ... and had no-one else to take care of them".[303] In such situations, demobilisation is virtually impossible without substitute care being provided – an unlikely event in the midst of conflict – and, by association and identification with the adult soldiers, children may be drawn progressively into assuming an active role in the conflict. When the children's primary source of protection is thus denied them, their situation poses a high degree of risk. It will be apparent, therefore, that in cases where factions forcibly recruit children, and in addition deny their participation, there is little likelihood that they will be

300 Case study for the the former Yugoslavia

301 From the case study for Sierra Leone prepared by the project Children Associated with the War (CAW) in: M. McCallin *The reintegration of young ex-combatants into civilian life.* ILO, Geneva (1995)

302 Case study for El Salvador

303 Case study for El Salvador. Note also the situation in Rwanda where some 5,000 children are being cared for by the RPA following the genocide of 1994. Here the military emphasised its protective role as the child's substitute family.

prepared to demobilize such a readily available source of manpower for any reasons associated with their welfare.

For those case studies from countries where conflict has ended, the situation is little better for the children. In no situation has there been formal recognition in peace agreements of the participation of children, even when their involvement has been extensively documented. In Mozambique, for example, where Renamo's recruitment of children was well known, "public recognition that children had been actively involved in the conflict was considered a political liability, and their demobilisation was inconvenient".[304] Although there has been no formal denial of the participation of child soldiers in Ethiopia, there have been no special efforts to address their situation in the demobilisation process. The large numbers of children and young people who were involved may have defied any attempt to target them as a matter of priority, although the education of some 8,500 is reportedly the responsibility of the Ministry of Defence.

In Cambodia, where again there has been no denial of the participation of child soldiers in the conflict, "there was no specific mention of child soldiers in the Paris Peace Accords of 1991. The demobilisation survey and proposal in 1992 does not deal with the problem or the specific issues surrounding possible demobilisation of child soldiers. It is implicit that they are seen in the same light as adults."[305] The proposal for the demobilisation of some 43,000 soldiers between 1996 and 1998, sponsored by the World Bank, includes no provision for child soldiers.

In South Africa, although there has been no formal demobilisation of young people involved in the self-defence/protection units, the normalisation of their daily lives is being addressed through a programme of re-education, retraining and vocational programmes, which is the responsibility of the Ministry of Welfare and Population Development. This process should be assisted by the process of disarmament which the government has implemented "by declaring an amnesty wherein all members of former units protecting communities could hand in arms in their possession".[306]

[304] M. McCallin (1995). *The reintegration of young ex-combatants into civilian life.* Report prepared for the International Labour Office, Geneva.

[305] Case study for Cambodia

[306] Case study for South Africa

In some instances[307] it was noted that material benefits in the form of cash, rations or clothing were awarded to former child combatants. In others, limited initiatives on the part of the government[308] and NGOs and other local organisations[309] are being implemented which seek to address the children's needs, notably education, vocational training and psychological counselling. Attention to former child combatants may also be integrated into programmes for children affected by war, such as the programme of family tracing implemented by UNICEF, Rädda Barnen and the ICRC for Sudanese unaccompanied minors, amongst whom were a significant number who had been directly involved in the fighting.

Whilst every effort must be made to secure their demobilisation, it should not be regarded as a discrete event, an end in itself, but more the initial step in rehabilitating and reintegrating the children. The fact that children's rights and welfare are either marginalised or completely ignored, even in official procedures for demobilisation, highlights the extreme vulnerability of children who are not only easily recruited, but can be summarily demobilised without due regard to their special needs. Thus the continued inadequacies of the international system in protecting their rights, and securing their well-being following demobilisation, must be emphasized.

These inadequacies are disturbing, given the emphasis many case-writers placed on the manner in which children have been affected. The experiences and events that characterize their involvement mean that, in a context of violent separation from their families and communities, they have lost their childhood and chance for education, whilst being "indoctrinated to obey orders without question" and "programmed to solve problems using a gun".[310] Recognition of their participation and its consequences for their

307 Burma/Myanmar; Ethiopia; Liberia

308 a) Mozambique: Programme of Support for Children with War Experience implemented in 1995 by the Ministry for the Co-ordination of Social Action.
b) South Africa: literacy and counselling programmes, and assistance with finding employment for 13–18 year olds involved in the SDU

309 a) Liberia: programmes of rehabilitation and vocational training
b) Nicaragua: therapeutic centres

310 Case study for Burma/Myanmar

development must be premised on a readiness to respond to their needs for rehabilitation and reintegration, and incorporated in peace agreements and related documents. Without this requirement, there is no framework within which planning for the demobilisation of child combatants can be undertaken, or programmes implemented which address their particular concerns. The limited initiatives described above underline the difficulties that can be encountered when such processes are undertaken on an *ad hoc* basis, without the support of policy and planning procedures.

THE CONTEXT FOR REHABILITATION AND REINTEGRATION

The significant rhetoric which has surrounded the phenomenon of the recruitment and participation of children in armed conflict, in particular the sensationalistic focus on "young killers", has hidden the necessity of recognising and addressing the essential child-welfare concerns which characterise every instance in which children are involved in armed conflict. There is almost a sense in some instances in which they are portrayed as being irretrievable; as though the acts they have committed put them outside the boundaries of civil society.

> *Even if they survive the rigors of combat, it's often too late to salvage their lives. Unrelenting warfare transforms them into preadolescent sociopaths, fluent in the language of violence but ignorant of the rudiments of living in a civil society.*[311]

Advocating for the demobilisation of child soldiers, and determining appropriate strategies for their rehabilitation and reintegration requires an understanding of the consequences they suffer *as children*, and a framework within which planning for children can be conceptualised and implemented. Planning for children in such extreme circumstances must incorporate an assessment of the risks to their development that have resulted from participation. It must be complemented by a consideration of the social, political and cultural reality which defines the context for the children's rehabilitation and social reintegration. It must therefore take

[311] Boy Soldiers. Newsweek Special Report, August 7th 1995

account of gender issues, the age of the children involved, and their length of service – a point of particular importance for those recruited at a young age who will have been separated from their families at a time when they "have not yet formed a stable identity which would allow them to psychologically 'survive' under such circumstances."[312]

The implicit needs/rights perspective of the Convention on the Rights of the Child provides a framework for guiding such an assessment through its consideration of the "evolving capacities" of the child, and the provisions of the various articles which ensure the child's right to grow and develop in an environment where her/his needs are met. The applicability of the CRC in the many and varied circumstances described in the case studies is a function of the extent to which it embodies a set of norms on how children should be treated, and represents a consensus on the minimum rights which must be guaranteed to children.

> *The text of the Convention on the Rights of the Child provides an excellent framework for advocacy on behalf of children, regardless of the discipline from which the rights of the child are being approached. It is equally useful to psychologists, educators, religious leaders, lawyers and parents ... Each time the standards of the Convention are invoked in practice, it strengthens its importance in international law.*[313]

Children are involved in an active process of development; they are not static observers of events. Thus whatever their situation, be it as the member of a functioning family, or a member of an armed faction, they continue to develop physically, emotionally and socially. Developmental outcome, or more simply their capacity to function effectively within their societies, is determined by their experiences in this process, be they positive or negative. The cheerful twelve year old who was abducted by an armed group

[312] E. Jareg. (1993) *Rehabilitation of Child Soldiers in Mozambique.* Note prepared for the Secretary General of Redd Barna

[313] C. Price-Cohen. *Considerations Affecting the Implementation of the United Nations Convention on the Rights of the Child in Situations of Forced Migration.* In: M. McCallin (Ed.) The Psychological Well-Being of Refugee Children. Research, Practice and Policy Issues. Second Edition (1996). International Catholic Child Bureau, Geneva.

may come home as an aggressive sixteen year old, carrying her own child in her arms, brutalised by abuse and with a confused sense of loyalties and identity.

The CRC, through its emphasis on the child's "best interests", addresses the heart of the issues that must be considered in implementing procedures to secure the well-being of child combatants, and requires us to assess whether and how our interventions meet this requirement.

Planning for the demobilisation of child soldiers must incorporate their special requirements for rehabilitation and social reintegration. It will, therefore, need to take account of the social, political and cultural context that resulted in their recruitment, the reality of their experiences in conflict, and their situation upon return. The experiences and conditions of life of their families and communities are integral components of each child's reality and, thus, efforts will need to be made to raise awareness within communities of the impact of the conflict on their children, and to incorporate their participation into programmes of rehabilitation and reintegration. In itself, this process can make a significant contribution to post-conflict resolution and the prevention of further recruitment. Greater understanding and awareness, and its corollary of greater empowerment, can prove crucial to preventing recruitment, as is evidenced by the disparity between the numbers of children from "educated" families who are recruited, as opposed to those from poor families.

Obviously, the above statements are easily if necessarily stated, whilst the reality is and will be harder to address given the circumstances of recruitment and participation. Several factors will influence the manner in which children's issues are incorporated into demobilisation procedures: the number of children involved, and in which faction; denial of participation/political sensitivity; demands on available resources, particularly in the period of post-conflict reconstruction; needs of other war-affected children. Without efforts to incorporate children's needs *generally* as a priority issue in the process of reconstruction, specific concerns for the situation of child soldiers may be marginalised.

> *... the State policy regarding the status of children in general remains that they are not "top priority". The government maintains that in the*

acute crisis situation which is alive in post-war Lebanon, "children's needs cannot be a priority".[314]

As with recruitment, attitudes towards the demobilisation, rehabilitation and reintegration of children may well reflect the low priority accorded to their well-being. As the case study for El Salvador notes, the hardships suffered during conflict are not necessarily any greater than those imposed by the poverty which characterises normal life, as evidenced by the ease with which children from marginalised and impoverished communities are recruited. In this context, the exploitation and marginalisation of their parents needs to be redressed, a decent standard of living and full protection of their social, political and human rights afforded them. The effect of such changes would quickly be reflected in the lives of their children.

In this regard it is worth noting the reservation outlined in the case study for the Philippines with respect to the term "reintegration".

> *Reintegration to what – a life which is dysfunctional, mainstream society that is saddled with social problems? What is indeed normal? The child soldiers learned poverty and injustices as normal realities of life – what they discovered though is an ideal when they joined the NPA. Therefore the use of recovery and not reintegration. For the child soldiers and rebel returnees, reintegration smacks of surrendering their principles and ideals to a society which in many ways is plagued with more problems and may be the very root of these problems.*[315]

This comment implicitly questions the extent to which initiatives directed at conflict resolution are also concerned to address the root causes of conflict, characterised to a large degree in the case studies by efforts to address social injustice and the social and economic marginalisation of particular ethnic or minority groups.

The case study for the Philippines also very importantly states that "the most appropriate policy and intervention can best be identified if the phenomenon of child soldiers is viewed from their eyes, their life situation, their

[314] Case study for the Lebanon
[315] Case study for the Philippines

context". With due regard to this comment, the following issues are outlined as of particular concern in planning programmes for the demobilisation, rehabilitation and social reintegration of former child soldiers.

THE FAMILY AND COMMUNITY

> *... the family, as the fundamental group of society and the natural environment for the growth and well-being of all its members and particularly children, should be afforded the necessary protection and assistance so that it can fully assume its responsibilities within the community, (and) that the child, for the full and harmonious development of his or her personality, should grow up in a family environment, in an atmosphere of happiness, love and understanding.*[316]

> *Children's well-being and development depend very much on the security of family relationships and a predictable environment. War, especially civil war, destroys homes, splinters communities and breaks down trust among people – undermining the very foundation of children's lives.*[317]

These two statements describe on the one hand the "ideal" to which all societies should aspire in protecting and caring for their children and, on the other, the "reality" that confronts the millions of children caught up in armed conflict. It is the gap between the two conditions that we seek to close when we consider interventions to address the well-being of children in conflict situations. The child soldiers are certainly amongst the most vulnerable of these children, not least because they have undergone their experiences whilst separated from their families and traditional support structures. Thus the principles of working with unaccompanied/separated children would seem to be those most applicable to their situation.

[316] Preamble to the UN Convention on the Rights of the Child.

[317] Promoting Psychosocial Well-Being among Children Affected by Armed Conflict and Displacement: *Principles and Approaches.* Save the Children Alliance (1996). Paper produced as a contribution to the UN Study on the Impact of Armed Conflict on Children.

It is not by chance, therefore, that the issues of family and community are given priority in this section. Perhaps the most significant factors influencing the child soldiers' return to civil society are family reunification and reintegration. It is here that the two contrasting aspects of their experience confront one another. The child and the soldier. The child has need of the family, and child soldiers, except in rare situations, have been separated from their families for long periods. Yet the soldier will need assistance to adjust and reintegrate, due largely to the process of "asocialisation" and "total destruction of trust in others" which will have been the experience of the vast majority of these young people.[318] All too often, it is the soldier we consider and we lose sight of the child. In recognition of the crucial role of the family in rehabilitation and social reintegration, planning must ensure that family tracing programmes are implemented as quickly as possible,[319] and children reunited with their families.

This will be an important process for the children, who may be confused by the situation, or indeed reluctant to relinquish their identity as a soldier, particularly when they perceive themselves as having played a role in addressing the ills that their society had experienced. This can result in a sense of abandonment or rejection following demobilisation, or resolution of the conflict. For many of them, the army or armed group will have become "their protector and provider, and they in turn (will have) identified with it. This loss, as much as the direct influence of militarisation, should be recognised as influencing their behaviour"[320] so that social reintegration is based upon a process of re-attachment to their families and communities.

Many children have been deliberately moved far from their families following recruitment.

... many were sent away to remote areas where contact with their home

[318] E. Jareg (1993) op. cit.

[319] For a full discussion of family tracing procedures see: L. Bonnerjea (1994). *Family Tracing. A Good Practice Guide.* Save the Children Development Manual No. 3. London, Save the Children.

[320] McCallin (1995) op. cit.

was difficult, impossible or forbidden. Some observers have speculated that distant field placements are chosen in order to make it difficult for soldiers to desert or defect ...[321]

Due to the process of the conflict families may be displaced, in some cases seeking refuge across a border, or in different areas within their own country. Displacement in such situations may occur several times as the fighting moves from one area to another. In the ensuing chaos, children may be separated from their families and recruited as unaccompanied children, or may join up because of the manner in which the conflict is prosecuted against their families and communities. In such circumstances, family tracing and reunification become very difficult.

The conflict in turn has also led to more children being displaced, orphaned or living in families where one or both parent/s are killed or the father is away fighting. Disruption of family units could also motivate a child to join the movement.[322]

Simply finding and placing the children with their families may not be sufficient, as the impact of the conflict on the family may mean that, on return, the children are confronted by a family that may have radically changed in structure and the manner in which it functions. This may significantly compromise the efforts the child soldiers and their families must make in order for them to return to life in civil society.

The simplest form of reunification is the direct return of a child to the members of the family she was living with before the separation. In the simplest reunification she is also returned to the same place they were living in before – displacement of the family was temporary and they are all back together ... the people are the same and the place is the same. But nevertheless the first impression of 'sameness' is illusory: the child has had some very frightening experiences. She has been without protection for a period of time. She may have been beaten, abused, raped. She is in many ways not the same child she was before. And the

[321] Case study for Burma/Myanmar
[322] Case study for Sri Lanka

family has changed too. Each member has had to develop survival strategies for themselves and for their relatives, sometimes with uncomfortable consequences. They may have had to choose which of their children should live and which should die. The mother may have had to turn to prostitution to survive. The father may have had to betray his community, his tribe, his religion, his caste in order to stay alive ... Reunification is not, therefore, a return to the situation that was before, but rebuilding, recreating a family, with new experiences which the family needs to come to terms with.[323]

In recognition of this, every effort must be made to assist families in this process of rebuilding family life. In Sierra Leone[324] this was accomplished through a programme of Family/Community Sensitisation, Mobilisation and Advocacy. This focused on:

- *promoting reconciliation and reunion among husbands and wives and settling of other family or inter-family or communal disputes; correcting family/community prejudices and misconceptions about child soldiers;*
- *animating and mobilising families and communities into actions on behalf of their children;*
- *creating readiness in families and communities for the return of their children back to the fold, even though times are hard and facilities for the care of children are inadequate;*
- *equipping families and communities with various skills in the handling of ex-combatant children.*

Such approaches have a greater chance of bringing about "a more positive social reality" through building "on a community's own social networks."[325] They are more cost-effective, culturally appropriate and sustainable, and reduce any likelihood that communities will become dependent on external resources to maintain a programme.

[323] L. Bonnerjea. op. cit.

[324] See case study for Sierra Leone in: M. McCallin (1995) op. cit.

[325] *Children in War. Community Strategies for Healing.* Save the Children USA, University of Zimbabwe & Duke University (1995)

There is an awareness that the Government itself will not be able to meet the many needs of the population. Thus there is a need to encourage community initiatives which are usually low cost and of greater impact, as well as to build up local capacity to resolve their own problems through programmes based on the communities themselves, managed by them and that are self-sufficient and sustainable.[326]

This comment also links with the observation made in the case study for the Lebanon that children's needs are not given priority during the phase of post-conflict reconstruction, when human and material resources are limited. Directing these scarce resources to the needs of one group of children – the child soldiers – risks marginalising or stigmatising them, thereby affecting their chances of reintegration. This will be the case particularly where a community blames the children for the events they experienced. Community involvement in their social reintegration would ensure not only the applicability of interventions to the circumstances of the children, but would also be more likely to result in an approach where the needs of the child soldiers were incorporated into overall programming to address children's issues in the period of post-war reconstruction.

Despite programmes of family tracing and reunification, however, there may be some children who will not return to their families. This may be because their families have perished in the conflict, or because they are fearful of returning home as a consequence of the events in which they participated, and may thus remain isolated from any system of care, protection and support. In some instances, even when the conflict has ended, the circumstances of the family may mean that it is "unable to satisfy the many needs of the child; it ends up 'facilitating' the move of the child onto the streets, either through the child's own initiative, or by encouragement or forced by the family to provide for him/herself and even for the family."[327]

For children in such situations there will be a need to consider alternative care-giving arrangements that meet the children's physical, social and emotional needs. This will be of particular importance for the older child

[326] Case study for Mozambique
[327] Case study for Mozambique

combatants who may have spent considerable time with the military and, as a result, find it very difficult to reintegrate into their families. As certain of the case studies pointed out, this group is in danger of resorting to criminal activities, made easier by the availability of weapons.

The creation of special institutions is not considered a satisfactory solution. They may serve only to marginalise the children even further, and also impede efforts at family tracing and reunification. They are also a costly enterprise which, though they may initially attract funding, are inappropriate in many cultures, and not a sustainable option for countries emerging from conflict. Nor are they the best setting in which to meet the special needs of children who have experienced horrifying and life-threatening events.

> *Children in institutions may get regular meals and exhibit playful behaviour but miss the individual love, nurture, guidance, role-modelling and personal attention that they depend upon for normal growth and development ... Institutionalisation leaves children disadvantaged, like plants that have not received the optimal nutrients for growth. Institution residents are often described as problem children, a judgement, if true, that reflects less about the children than about the failure of the society to meet their needs.*[328]

There is some concern that former child soldiers may need a period of time between demobilisation and return to their families to allow for rehabilitation and assessment of their situation. The acts they committed may have had a serious impact upon them, they may be in poor physical health or have become dependent upon the drugs that were administered to them. What is important to recognise here is that any such programme of rehabilitation does not become an end in itself. It must be seen as a temporary, interim measure. The message must be given to the children and their families that this is not the answer to their situation, but a process which is designed to facilitate their ultimate return to their families and commun-

[328] E. Ressler, J.M. Tortorici & A. Marcelino (1993) *Children in War. A Guide to the Provision of Services.* UNICEF Publications.

ities. Such responses will also be a function of the number of child soldiers involved in the conflict, and the availability of resources. The programme in Sierra Leone which implemented such a response was working initially with a small number of 370 children. In countries such as Ethiopia, where the numbers of child soldiers ran into thousands, it would have been an impossible task to implement a programme of this nature. Thus, some thought must be given to this issue when planning the demobilisation of children, in particular preparing communities for the direct return of their children so that rehabilitation can be integrated into community activities.

A more appropriate option may be placement with a substitute family. This "may be a welcome relief to some children, who had been previously fending for themselves" but for others "substitute families may be met by fear, anger and resentment."[329] The motivation of substitute families must obviously be a prime consideration in order to protect children from abuse and exploitation. But the willingness of many families, even in impoverished circumstances, to take in an unaccompanied child, and their capacity to nurture the child has been recognised in many situations. In addition to the increased benefit to the children, such arrangements are less costly, and can direct resources to local communities, thereby strengthening their own capacities to care for their children.

Efforts at family reunification may also be influenced by attitudes within the military. They may not wish to admit to numbers of children in their forces, so they conceal the issue, or may consider that children can be better helped by remaining the responsibility of the military.[330] In the latter

[329] See, for example, Bonnerjea (1994) op. cit for a discussion of factors to be considered in the placement of children with substitute families. Also the experience of Rädda Barnen described by U. Blomqvist, *Social Work in Refugee Emergencies. Capacity Building and Social Mobilisation: The Rwanda Experience.* In: M. McCallin (Ed.) (1996) op.

[330] For example, the NPA in Uganda (see case study for Uganda in M. McCallin (1995) op. cit.), and Rwanda where a special school has been established for children associated with the Rwandan Patriotic Army in the course of the genocide of 1994, rather than effecting procedures for immediate family tracing and reunification. (Secondary source)

situation, evidence indicates that in many instances the army gathered up unaccompanied children, at least initially as a 'protective' measure. Here, child welfare issues, in particular the importance of family life for the children, may need to be introduced in a sensitive manner with the military in order to involve them in the children's reintegration. Insistence on the immediate removal of the children could be interpreted as in some way negating their concern for the children's welfare. But where the military has shown such concern, this can be built upon, and at least their knowledge of where children were found could assist in family tracing and reunification.

ECONOMIC CONSIDERATIONS

To a disproportionate extent, child soldiers come from poor and/or marginalised families. It is these communities which undoubtedly suffer the worst deprivations as a result of conflict, and face increased impoverishment as the conflict continues.

> *All parents are losing their sons, so families don't have enough labor to support themselves any more. Every family is getting poorer and poorer ... It's as though the SLORC is strangling the people to death. 55 or 60 boys already have to go out of every 100 households in my area.*[331]

Economic concerns can influence the process of demobilisation of children, even when families do not support the participation of their children. In Afghanistan at present, despite the fact that "people hate the war", there are no alternative sources of employment, and involvement in the conflict is a source of income.

The capacity for communities to accomplish economic recovery can have significant implications for those children who participated in the conflict. The need to contribute to the family's well-being can affect, for example, their ability to take advantage of even the limited opportunities for educational or vocational enhancement that are available.

> *Due to the heavy human loss in Afghanistan a large number of children*

[331] Burma/Myanmar case study

are now the earning members and have to work hard to support their families.[332]

Also, the time spent with the military may mean that the "child soldier" is actually a young adult when she/he returns home. The effect of long periods of time with the military may be that children and young people find themselves "alienated from the traditional means of livelihood available for them"[333] as they have never had the opportunity to learn to do agricultural work or obtain the skills which underpin the economic basis of their communities. Here there is an obvious link between economic considerations and reintegrating into family life and civil society. For most communities, "a concern for broader social and economic reconstruction (is) the most significant factor in the remaking of their worlds."[334] Allied to this is the extent to which children will be "valued for their contribution to the productive work of the family."[335] Thus where children can assist their families in economic recovery, the process of reintegration may run more smoothly. The children's capacity and willingness to work will be of immediate and appreciable value and will contribute to their re-attachment to the family and community, as well as being an "opportunity to 'deconstruct' the past", and to "develop an alternative meaning to their lives"[336] that is not associated with the power wielded by the gun.

EDUCATION AND VOCATIONAL TRAINING

In general, time spent with the military is seen as a time of lost opportunity. The theme of lack of skills learned in the military was pervasive, and examples of efforts to provide even limited education during conflict were few.[337]

I didn't receive any education apart from basic training ... I don't think

[332] Case study for Afghanistan.

[333] Case study for El Salvador

[334] S. Gibbs. *Post-War Social Reconstruction in Mozambique: Reframing Children's Experience of Trauma and Healing.* Disasters Vol. 18 No. 3 (1994)

[335] Gibbs op. cit.

[336] McCallin (1995)

[337] El Salvador, Burmese opposition forces, PKK

there are any post-service educational opportunities for child soldiers, because by the time you get out you're too old.[338]

Not through any fault of their own the adjustment they have to make is difficult. Unemployment is high among former members of the self-defence units. Their lack of education and training is particularly debilitating. The dislocation of the educational system does not help them in any way and many teachers are not equipped to deal with the difficulties they experienced and exhibit.[339]

This last reference touches on two issues that can have an impact on the child soldier's rehabilitation and reintegration. Not only have the children received no or limited education as soldiers, but in many countries the conflict destroys those institutions which serve children – in this case education – making it more difficult for them to resume normal life. Also, the individuals whose role is to assist children, here the teachers, are themselves affected, and may find it difficult to make the effort to assist children with special needs. There is obviously here a requirement to target the reconstruction of the educational system as a priority, not only for former child soldiers, but all children, and to ensure that teachers are given support and, where necessary, additional training to cope with the added demands of teaching children who have been affected by conflict.

(The teachers) are aware that many have problems concentrating and are accepting and understanding if a child becomes restless or needs to leave class for a break. In conversation with the children, usually the first thing they want to show a visitor is their school books. They see education as their way out of poverty and are thus motivated to learn ... The activity of going to school ... is clearly having a major organising effect on the children's behaviour and self-esteem ... It is important, therefore, that the teachers are given the assistance they need to be able to teach.[340]

338 Burma/Myanmar case study

339 Case study for South Africa

340 Jareg, E. & McCallin, M. (1993) *The Rehabilitation of Former Child Soldiers.* Report of a Training Workshop for Caregivers of Demobilised Child Soldiers. Freetown, Sierra Leone. Geneva, International Catholic Child Bureau.

Given their limited educational opportunities, and the need in almost all situations to contribute to the family's economy, "school" for many former child combatants will likely require a vocational component to assist the children to acquire skills and find employment.[341] A number of the case studies made reference to the need to establish a secure economic base after conflict, and how this can influence the children's capacity to benefit from educational and vocational programmes.[342] To secure effective reintegration, and in some cases prevent re-recruitment or criminal behaviours, it will be very important to recognise the inextricable link between economic and educational concerns. Children miss out on education because of their participation, leaving them with few or no skills to secure employment and thus further marginalised. They are under pressure to contribute to the family's economy and, where participation in the conflict remains an option, will be unlikely to relinquish this source of income.

> *A large proportion of child soldiers in Afghanistan are illiterate. Since their present age is not appropriate to pursue education, they still prefer to be involved in the current conflict. It will take a long time to educate and encourage child soldiers to cease their involvement in the conflict. They need to be encouraged and provided with a special educational programme and other activities ... They should also be provided with alternate employment during the course of their studies to support their families.*[343]

Not wanting to return to school because there will be too much effort involved in catching up on lost time, or because they may be targets of abuse or victimization because of their involvement,[344] can affect the

[341] This issue is one which will be dependent on the age of the children, as many (at least in theory) will be too young to enter the employment market. See M. McCallin (1995) for a discussion of formal, non-formal and traditional apprenticeship training schemes as a means of combining vocational training with the children's need for basic education.

[342] See case study for El Salvador where only 6% and 32% of those eligible for basic and vocational education programmes respectively ectually enrolled.

[343] Case study for Afghanistan

[344] Uganda case study

extent to which former child solders take advantage of educational opportunities.

The importance of re-establishing educational systems must be noted because they provide an environment where children can gain "cultural literacy" and which "allows children to gain a sense of social responsibility".[345] Whilst these important aspects of education were noted in the context of on-going conflict, they are crucial components of effective rehabilitation and reintegration. Where communities are striving to normalise daily life "the organised structure of schools also helps to provide children with a sense of security essential to healthy child development".[346] It sends a positive message to children that their needs and welfare are matters of concern. Schools can also be the place where children develop "greater feelings of empowerment for strength and competence". They are also the natural setting for peace education. "Resolution" of a conflict does not imply that peace automatically breaks out, and violence may be a continuing characteristic of the children's lives.[347]

Restoring and, where necessary, restructuring educational opportunities for child soldiers will be a function of local conditions, and the ages, numbers and gender of the children recruited and the functions they performed. Whatever the process followed, there are alternative methods to traditional educational structures and systems that will need to be considered, including retraining of teachers to address special needs that arise due to participation in conflict.

CULTURAL AND SOCIAL ISSUES

The breakdown of traditional family and community structures as a result of conflict and in some cases the resulting empowerment of children, as happened for example during the Intifada and the struggle against apart-

345 Case study for the Intifada

346 Case study for the Intifada

347 See, for example, the case study for South Africa where "Violence based on ethnicity was...encouraged by the state" and there is now a need to reconcile "the different warring parties." Also, for a description of a programme to introduce peace education in primary schools, see: J. M. Tortorici. *Peace Education in Nicaragua*. In: M. McCallin (Ed.) (1996) op. cit.

heid in South Africa, can have profound implications for the children's rehabilitation and reintegration. Parents are rendered powerless to protect their children, often with the deliberate intention of undermining traditional values, and creating tension and discord within families. Post-conflict it appears unlikely that children who had engaged in activities beyond their years will readily submit to the old ways, and communities will need to be enabled to incorporate what may be lasting changes to their social and cultural systems.

> *Throughout the country the youth were exhorted to become young lions in order to shake off the yoke of their oppression. Black youths became accustomed to the notion of violently engaging in conflict with forces of the state ... Violence became a socially sanctioned means of attaining change within black urban populations.*[348]

In many conflicts, families and communities themselves encourage children from an early age to incorporate beliefs and standards that protect their cultural and ethnic identity. Once the conflict has ended, they may be reluctant to submit to restraints upon their behaviour, seeing themselves as the prime movers in effecting the desired social and political change.

> *The long-running civil war has meant that military service is something of a legacy, passed on from generation to generation ... Children who grow up in communities ... which face frequent attack and relocation develop political sensitivity against the central government at a very young age. They are raised to associate military induction with the survival of their ethnic culture.*[349]

Contrasting with such situations are those where children from certain ethnic or minority groups are targeted with the purpose of destroying their social structure. Here many young people may be permanently alienated from their communities of origin.

> *... forcing under 18s (from indigenous communities) to enrol in the army*

[348] S. Mokwena, *Marginalisation, Youth and Violence.* Quoted in case study for South Africa

[349] Burma/Myanmar case study

practically severs and destroys the future of these communities ... the process of acculturation, side-lining and dehumanisation that the youths are subjected to signifies a form of ethnocide. For these youths lose their identity, being ashamed of their origins and ancestral roots, often not returning to their communities on leaving military service.[350]

Even when this process is not deliberate, it must be recognised that for the children the separation from their families implied by their participation in conflict means that they lose "access to social and cultural values, which is even more serious given that (they) were in the crucial stages of development."[351]

Communities must be enabled to recognize the long-term consequences of the induction of their children into "... a political culture defined by the idea that established authority is illegitimate ..." in order that their children may realistically benefit from the opportunities available in the period of post-conflict resolution, and not be hindered because they "can't suddenly discover the virtues of co-operation merely because the nature of the authority has changed".[352]

The contrary situation to this is one where children are considered difficult because they lack initiative "having been under military discipline for a long time, and having been provided with all basic necessities".[353]

During military service, the children are separated from the supportive and nurturing environment of the family, and are dependent on rigid and authoritarian structures to control their behaviour ... This is the way of life they know, and the one on which they model their own behaviour towards others. Also they were unused to making decisions for themselves and may initially exhibit a continued dependency on authoritarian structures to control and limit their behaviour. Once this is removed, and without a support structure in place, the risk may be that not having learned accepted mechanisms for personal control they may resort to aggressive techniques to get what they want.[354]

[350] Case study for Guatemala

[351] See case study 2 for Mozambique in M. McCallin (1995) op. cit.

[352] Case study for the Intifada

[353] Case study for El Salvador

[354] M. McCallin (1995) op. cit.

A number of the case studies made reference to negative attitudes within their communities as a result of the children's difficult, violent or even criminal behaviour. Expectations of a smooth and trouble-free progress towards social reintegration will in many cases be unrealistic. The cumulative effect of their experiences as soldiers, their separation from their families, and the likelihood of inadequate provision of basic services upon return will mean that, for many, social reintegration is the most difficult phase they encounter in a process fraught with risks to their well-being.

> *More (disturbing) ... is the fact that they are being seen as the persons responsible for the crime wave. This is not so says a former unit member ... There are some members but then their behaviour stems from other factors and not merely because they are members of the SDUs. One of these is the fact that many former members ... find it difficult to return to school.*[355]

One of the most difficult issues that some communities will confront when children return home is whether and how children are held responsible for the acts they committed as soldiers. Recent work by Save the Children (USA) in Rwanda indicates that "the application of *legal justice* might not serve the requirements of the *social justice* expected at the popular level".[356] Whilst this work may be considered to describe an extreme case, reflecting conceptions of the culpability of children who participated in acts of genocide, it has applicability particularly in those situations where there is an ethnic or tribal dimension to the conflict, or where formerly opposed factions are now required once again to live side by side. Because we know so little about what happens to children after demobilisation, the issue of juvenile justice as a factor influencing social reintegration was not documented in the research. It requires further study to ensure the protection of child combatants following demobilisation, and to promote post-conflict reconciliation between opposing groups.

[355] Case study for South Africa

[356] *Children, Genocide and Justice. Rwandan Perspectives on Culpability and Punishment for Children Convicted of Crimes Associated with Genocide.* Save the Children Federation (USA). Kigali, 1996.

> *They tend to be aloof from others with little confidence in themselves and distrustful of others. They tend to lapse into absentmindedness as well as swift mood changes for no apparent reason ... Some ... are reported to be undisciplined and difficult to control by parents or guardians. One such group of boys has acquired guns and is laying ambushes on passers-by ...*[357]

> *There is a huge difference between the educational level, skills, behaviour and way of thinking of a child soldier and a child not directly involved in armed conflicts.*[358]

The physical and emotional well-being of child soldiers at the time of demobilisation and reintegration is influenced not only by the experiences inherent in direct participation in conflict, but also the manner in which they as individuals were treated by the factions which recruited them. The impact of "...physical and emotional abuse... used to reiterate the importance of rank and subordination, involving debasement and humiliation of the younger recruits",[359] can result in low self-esteem, guilt feelings and violent solutions to problems.[360] Lack of trust and confidence in others, and aggressive or militaristic behaviour may be the real or supposed consequences of such experiences. The latter is often the preferred, or at least presumed, effect of children's direct participation in conflict, as evidenced by attitudes of people involved in the rehabilitation of former child soldiers in Nicaragua:

> *Until that moment we only knew what was said about them: "bloodthirsty children", "humantigers", "they take out people's eyes", "born assassins" ... According to the first director of the project, the personnel had even been warned of the need to prepare themselves physically by taking karate or self-defence lessons so as to be prepared for any possible aggression.*

[357] Case study for Uganda
[358] Case study for Afghanistan
[359] Burma/Myanmar case study
[360] Uganda case study

An intoxicated young rebel with the United Liberation Movement of Liberia (ULIMO) in 1996. In many wars, child soldiers are supplied with drugs routinely or before battles. Photo: Reportagebild

Reintegration is seen as less of a challenge where socialisation within an opposition movement has given rise to a sense of empowerment, and where "the movement too becomes family to the children".

> *... child soldiers are treated with respect and as human beings ... benefits are equal to or translated to mean self-improvement, self-esteem and confidence, respect, the opportunity to serve the masses, the poor.*[361]

In this regard, the meaning of the events which the children have experienced, and their understanding of these events, can act as a buffer to reduce negative consequences for their emotional well-being. One theme which did emerge was particular to certain rebel or opposition forces who did attempt to give some meaning to the children's participation – an awareness of fighting for democracy, a sense of shared purpose and belonging – "the relationship and social welfare that I ... experienced ... will be useful".[362]

> *... an ideological explanation of violent conflict, along with a sense of being on the "right side", helped to buffer Palestinian children against some of the negative psychological effects of violence. Such a cognitive framework was important in helping children work through their reactions to violence and gain a sense of competence and mastery.*[363]

In most cases the physical well-being of child soldiers is seriously compromised, due to inadequate nutrition and health care, and the effect of harsh punishments[364] and training regimes. Children suffer from worm infestation, respiratory tract infections and water-borne diseases due to poor hygienic conditions and lack of basic facilities. Many are known to be suffering from sexually transmitted diseases, and it may be supposed that this

[361] Case study for the Philippines

[362] Burma/Myanmar case study

[363] a) Baker and b) Anthony & Cohler in case study for the Intifada

[364] In Liberia a form of torture known as 'tabay' was practised on children. This involves tying the elbows together behind the back for prolonged periods. The result is nerve damage to the arms. Cited in: *Easy Prey. Childs Soldiers in Liberia.* Human Rights Watch/Africa & Human Rights Watch Children's Rights Project. (1994)

problem will be common to almost all cases due to the prevalence of sexual abuse of child combatants reported in the case studies. The routine administration of drugs and alcohol to enhance the "performance" of child soldiers in combat[365] will also have an impact on their physical and psychological well-being. The long-term implications of substance abuse, where "... young soldiers were given amphetamines, tranquilizers and alcohol before being sent into battle"[366] are considerable, particularly when the availability of such substances after demobilisation may compromise the process of rehabilitation and reintegration.

Of particular concern when one considers social reintegration will be the extent of physical disabilities that child soldiers suffer. Previous sections have described the appalling injuries inflicted on children. Their disabilities will in fact pose additional problems for them, and may result in their marginalisation or rejection by their families and communities who are unable to cope with the additional burden of a disabled family member.

> *N was 14 when he became a child soldier. Having lost his hearing and two hands in combat, married with two children, he finds himself ... with no work and no resources. "Nobody will give work to a former militiaman, we are avoided like the plague. Yesterday I had to stop sending my children to school. I did not have any money to pay for their transport."* [367]

Sexual abuse of the children which, in some case studies, is reported to be widespread, will have a particularly serious impact on the capacity of girl soldiers to reintegrate with their communities, as their families may be unable to accept or acknowledge their use as "wives" or "comforters" of the male combatants.

> *Girl children are more vulnerable to sexual abuse than adult women or maturer girls who are shunned as having been already "used" by men and who probably could have HIV/AIDS already ... In some instances*

[365] Liberia

[366] Burma/Myanmar case study

[367] Case study for the Lebanon

15-year-old boys have been allocated girls as wives. Such boys become proud and happy though they are likely to end up with sexually transmitted infections.[368]

... there were girls who were obliged to have sexual relations with the combatants ... The women among other things had to "alleviate the sadness of the combatants." And who alleviated our sadness after going with someone we hardly knew? ... At my young age I experienced abortion. It was not my decision, I could not decide on that. They decided, in any case. Hadn't I handed over my entire life? Had I not undertaken a commitment to permanent obedience and discipline? ... There is a great pain in my being when I recall all these things, principally because with time I have come to understand that to be a woman in any group was always a disadvantage. In spite of my commitment, they abused me, they trampled my human dignity. And above all, they did not understand that I was a child and that I had rights.[369]

Cultural beliefs and attitudes may affect the girls' opportunities for marriage, with the result that they do not stay with their families and see prostitution as their only means of livelihood. Additionally, communities may find it difficult to accept that their daughters were also combatants. Many young women, whose identity was forged in combat, may have to conceal the role they played in conflict and, "for fear of total rejection by their husband's family (must) pretend to be the gentle, soft-spoken and submissive woman that their civilian counterpart is."[370]

The issue of long-term impacts of sexual abuse of the boys is one about which little is known, and which will require further sensitive enquiry, in order to take account of cultural mechanisms to deal with such a delicate problem.

[368] Case study for Uganda

[369] Former female combatant, interviewed for the case study for Honduras

[370] P. Maramba, (1995) *Reintegration of Demobilized Combatants into Social and Economic Life. The Zimbabwean Experience.* Report on a consultancy for the ILO, Geneva.

TRAUMA AND RECOVERY

It will be apparent that the social reintegration of child soldiers into families and communities devastated by the effects of conflict will be a far from straightforward process. In particular there is a danger that a focus on their "traumatic" experiences, and the psychological consequences of these experiences, may lead to the implementation of a Western therapeutic or treatment model to address their situation. As with other groups of children made vulnerable because of armed conflict, there is a number of considerations that affect such an approach. The arguments against such a model focus on its cultural inappropriateness, and the resulting marginalisation or even "misdiagnosis" of many children. Other concerns point to the fact that it does not incorporate the knowledge and expertise of the local community, is costly to implement, reaches only a small proportion of the population, and contributes little to building capacities in local communities.

The alternative strategy, whilst recognising the impact of traumatic events, considers that children's responses may be influenced as much, if not more so, by their experiences of family loss and separation, and the precarious nature of family life that results from conflict.[371] Thus approaches to healing and recovery focus on re-establishing nurturance and support from within the child's immediate environment of home, family and community.[372] There is now considerable experience available on community development approaches with war-affected and displaced people which indicates that they provide a more flexible response that takes account of the social context, and incorporates the community in the process of defining and implementing solutions to their situation.

371 For a discussion of "secondary stress factors" that children may experience see: E. Jareg, & P. Jareg, (1994) *Reaching Children through Dialogue*. McMillan Press.

372 Significant expertise has been developed in recent years within international organisations such as UNICEF, and local and international NGOs on the implementation of community-based approaches which is relevant to the social reintegration of child soldiers. See, for example, the bibliography in the Save the Children Alliance paper mentioned above. Also, S. Fozzard, (1995). *Surviving Violence. A Recovery Programme for Children and Families.* Geneva, International Catholic Child Bureau.

Typical characteristics of community development approaches are:

- an emphasis on the need for *long-term development* and not just the more immediate relief of suffering;
- a strong emphasis on the importance of the *social context* in which people experience stress, and not just individual suffering;
- seeking *community definitions of needs and priorities* rather than professional judgements about them;
- seeing *traumatic experience as one among many aspects of stress* faced by people: distressing experiences of violence may be compounded by current difficulties such as poverty, continuing fear, poor housing, unemployment, lack of social resources;
- emphasis is placed both on the *collective nature of problems* and the need to promote *collective coping mechanisms.* It sees communities as resourceful and potentially powerful;
- community-oriented programmes tend to focus on *common needs within the community* rather than the specific needs of individuals.[373]

Over and above theoretical considerations of the relative value of these contrasting approaches, generally conducted in the sanitised environment of conferences in North America and Western Europe, it is the near-universal acceptance of the Convention on the Rights of the Child which, in effect, defines the applicability of community-based interventions. This is most evident in the "best interests" provisions of the CRC.

> *If, according to the best interests of the child, children and their families are positioned at the centre of all strategies to close the gap between reality and the optimum environment for full development, then a model*

[373] Reproduced from: D. Tolfree, (1996). Restoring Playfulness. Different Approaches to Assisting Children who are Psychologically Affected by War or Displacement. Rädda Barnen, Stockholm.

that treats symptoms rather than empowers whole human beings may be inadequate and counterproductive ... (T)he medical model ... tends to universalize a Western notion of child development which, even by using the term trauma, pathologizes children's invisible wounds and views them only as passive victims rather than active survivors.[374]

To respond to the needs of the children from the limited perspective of a trauma or sickness model is to deny the complexity of their situation, and a failure to recognise that the abuse of children as soldiers is, in almost all instances, a consequence of poverty, repression and social injustice. More than psychotherapeutic interventions, their situation and that of their families cries out for interventions that improve their conditions of life, and acknowledge the role of the family and community in effecting the recovery of children who have experienced distressing events.

Within such a framework, international organisations, rather than co-opting to themselves the role of the principal actors in developing programmes of assistance and intervention, can focus on facilitating attention to the needs of the children through dialogue, partnership and advocacy with affected communities. In this way the "spirit" of the Convention on the Rights of the Child is translated into practical realities that promote the well-being of the children. Rather than a remote instrument of international humanitarian law, its implementation at local and national levels, both as a tool in advocacy and as the reference point for guiding assessment of the situation of children, ensures that children benefit from adherence to its provisions.

The above issues principally describe child welfare concerns that influence the social reintegration of child soldiers. It is important to recognise, however, that they cannot be addressed simply from a child welfare perspective implemented on an *ad hoc* basis in each situation. They require more determined advocacy to implement the overall provisions of the

[374] For a full discussion of the relevance of the CRC to this issue see: D. Reichenberg, & S. Friedman, *Traumatized Children. Healing the Invisible Wounds of Children in War: A Rights Apporach.* In: Y. Danieli, N. S. Rodley, & L. Weisaeth, (Eds.) International Responses to Traumatic Stress. Amityville, New York, Baywood Publishing Company, Inc. 1996

Convention on the Rights of the Child in order to ensure "the protection and harmonious development"[375] of the children. The issues are complex and interrelated, and require a political will at both local and international level to address the conditions of life that not only lead to recruitment, but also compromise the children's rehabilitation and social reintegration. We need to create a more informed understanding not only of the social and political conditions that give rise to recruitment, but also of the consequences for children's participation in conflict, in order to facilitate attention to their needs and protect them from further harm.

One aspect of protection in this context is to raise awareness that there are ethical issues to be considered in achieving a balance between directing attention to the plight of the children, without exploiting any individual child. All too often, interest in child soldiers focuses on the sensationalistic nature of their involvement in the conflict – the acts they committed – with little explanation of the circumstances that lead to recruitment or the consequences of participation for the children's well-being.

> *We often come across children who have been pressured into telling and re-telling their "horror stories" to journalists, researchers and sometimes even to officials of aid agencies and psychologists ... Field staff cannot avoid the need to collaborate with researchers and journalists: it is the terms of such collaboration that require clarification.*[376]

This "retelling" of his story is not good child welfare practice. In the course of a visit by one of the authors to the school in Rwanda that has been established for children associated with the RPA, the Director expressed concern about the number of visits to the school by journalists and representatives of foreign organisations. Children had been photographed and questioned about their experiences. They were often left distressed as a result, and did not understand why there was no follow-up to these visits. Repeated questioning about events in which he participated may compro-

[375] Preamble to the UN Convention on the Rights of the Child

[376] op. cit. Paper prepared by Save the Children Alliance (1996). This paper also presents guidelines on interviewing children.

mise a child's rehabilitation and reintegration as he sees that he is valued only with regard to his identity as a soldier, or worse, a "killer", and at times may lead to children "embellishing" what are already horrific events to gain attention. The child's physical security may also be put at risk, if he can be readily identified from photographs or other details about his background. Such situations infringe the child's right to privacy, and his right for recovery to take place "in an environment which fosters the health, self-respect and dignity of the child."[377]

[377] Articles 16 & 39 of the Convention on the Rights of the Child

Islamic Jihad holds a memorial for a Palestinian killed in a suicide bomb attack. Suicide missions are the particular province of adolescents.
Photo: Gamma/IBL.

CHAPTER 6

Military Attitudes to Child Soldiers

A DIFFERENCE BETWEEN FORCED recruits and volunteers which is often merely technical but sometimes can be quite significant is that the volunteers are not picked out by the military but present themselves for recruitment either spontaneously or as a result of active canvassing for recruits. In either case there is the opportunity for the military to vet those who present themselves, with a greater or lesser degree of selectivity.

In some instances, there is no evidence of volunteers being turned away as too young, or for any other reason. In others there is at least a nominal adherence to minimal recruiting ages, and some specific instances of refusal or diversion are given in the section on preventive strategies below.

Various reasons are given for the positive choice to recruit children, either forcibly or as volunteers:

> *... children are much more easily moulded than adults, who are already "corrupted by the ideas of capitalism".*[378]

> *They can easily be trained and disciplined, have a high degree of camouflagibility, and can run risky missions. The most significant factor however is [that they are] easy to mobilise whether recruited forcibly or voluntarily.*[379]

> *... children are easier to intimidate and scare with punishments and*

[378] Case study for Peru (translated from the Spanish)

[379] Case study for for Ethiopia

> *they obey more easily. Also, children are less inclined to desert than adults. As children have few opportunities outside the barracks, especially in relation to work, it is easier for them to become adapted to their new life conditions, rather than an adult recruit taken from his work, studies or family.*[380]

> *... the young fighters normally have no family responsibilities and are healthy. They are also easily controlled and they naturally obey adults.*[381]

> *Some ... military commanders claim that child soldiers are more obedient, do not question orders, and are easier to manipulate than adult soldiers.*[382]

> *Children are an attractive option for guerilla leaders as they are easily conditioned and motivated, and can be put into battle with less training. Although the children are put into more danger, they too are of greater danger to their adversaries.*[383]

In some contexts, as previously discussed, children, especially the youngest recruits, are seen as having a particular value which mature soldiers do not possess, as spies, decoys, lookouts, minesweepers, and so on, while teenagers are recruited and trained specifically for assassinations and suicide attacks.[384]

There are, however, opposing viewpoints. Not all commanders share the assessment that children can be tractable troops with particular advantages in certain circumstances. For some, their employment is morally repugnant. For others, they are seen even in purely military terms as an encumbrance. It is noteworthy that the only case study to include the results of widespread interviews with (opposition) commanders revealed a number of dissenting opinions:

[380] Case study for Paraguay (translated from the Spanish)
[381] Case study for Liberia
[382] Case study for Burma/Myanmar
[383] Case study for Sri Lanka
[384] Case studies for Lebanon and Peru

I try as much as possible to avoid having the really young ones in my unit. They just can't carry the heavy artillery. Kids slow us down. We like to keep them in the kitchen, or doing chores back at base.

[Some commanders] state that child soldiers are still children, and that they often misbehave, are unruly, and require frequent scolding, as young boys usually do. One ... Captain claimed that child soldiers often had to be taught the most basic life skills, such as how to wash, cook and take care of themselves.

Children are mostly good fighters, but they are not always careful ... When there is shelling, the younger ones forget to take cover. They get too excited. They have to be ordered to get down inside the bunkers.

Both the lifestyle and the duties of military service are extremely demanding and, with a limited force of soldiers and arms, ability to perform is critically important.

Some really young children ... are not ready or suitable for this kind of work. I send the ones who are not ready to school here, and the army funds their schooling. As for the children who do not seem cut out for this work, I arrange for one of the older boys to plan a "desertion" or escape as they will not leave without a leader.

In my regiment, there are some 12 and 13 year olds. I try to send them to school if I can; then they can join and become a soldier later if they decide to, or return to their families.[385]

These are not the only armed opposition groups opposing the involvement of children:

[The] policy of the liberation movement was actively to discourage youth, and in many instances including those over 18, from joining [their armed wings]. A direct result of this policy ... was the establishment of the ... College of Education in 1978 to educate the youth ... Persons under the age of 18 years who wished to form part of the armed

[385] Case study for Burma/Myanmar

wing were actively counselled against joining the armed wing and rather encouraged to continue their schooling or studies.[386]

Furthermore, the attitude of commanders may change with experience of the difficulties of controlling children with access to arms and bomb-making equipment, and the greater likelihood of them succumbing if caught:

The refinements of [government] interrogation techniques was another factor which forced [the armed opposition group] to rethink on the role of the [youth wing]. Youths appeared more susceptible to interrogation. The result was that in 1976 numbers in prison rose dramatically; a high percentage of these were youths.[387]

As children they are prone to adventurism and abuses. They tend to abuse the use of guns. The young rebel surrenderees, the ones who benefitted from a weak political foundation, are now into illegal gambling.[388]

Some professional soldiers share these reservations about the participation of children:

Leaders of the coup [of high-ranking officers] stated their reasons for the coup attempt as mainly being the stalemate in the armed conflict and the increased involvement in the armed forces of children whom they see as immature and negatively bearing on the morale and effectiveness of the armed forces.

... members of the regular force are often heard attributing the defeats and unruliness of the [government] army to the fact of recruitment and service of children ...[389]

[386] Case study for South Africa
[387] Case study for Northern Ireland
[388] Case study for the Philippines
[389] Both from the case study for Ethiopia

CHAPTER 7

Legal Standards on Recruitment and Participation

THERE ARE TWO TIERS of applicable law: national law and international law. Most national law on recruitment is clear and straightforward, and predominantly sets 18 years as the minimum age for recruitment, although there are a few exceptions with lower or higher minimum ages and somewhat more that permit volunteers below the age of 18.[390] Even in these cases, some governments have a policy of not sending under-18s into combat.[391]

However, there are too many instances in which the national law is ineffective. One problem is that, in the world today, armed conflict is almost exclusively in the form of internal rather than international armed conflicts. This highlights the practical limitations of national legislation on recruitment. When a government regulates recruitment into armed forces, it is only addressing recruitment into its own armed forces. This means that recruitment into armed opposition groups (which are, by definition, illegal in any case), is not covered by the law, nor are the military regulations penalising illegal recruitment applicable to those recruiting for such groups. When legislating, governments do not normally contemplate the possibility of civil war (other than by creating offences such as treason)

[390] See Table 2

[391] For example, Australia and the Netherlands

and, therefore, do not attempt to regulate recruitment by armed opposition groups, for example through the criminal law.

With respect to international law, two branches are relevant to the issue of recruitment of children into armed forces ("recruitment" covers both compulsory and voluntary enrolment)[392]: international humanitarian law – the law which governs the conduct of armed conflicts – and human rights law. At present, three international legal instruments govern the recruitment of children into armed forces and armed opposition groups and their participation in hostilities ("hostilities" means a battle or similar armed engagement in an armed conflict): Additional Protocols I and II to the Geneva Conventions of 1949, and the Convention on the Rights of the Child.

The Convention on the Rights of the Child was adopted by the UN General Assembly in 1989 and nearly every country in the world is now a party to it.[393] Article 38 of the Convention provides, in part:

> *(1) States Parties undertake to respect and to ensure respect for rules of international humanitarian law applicable to them in armed conflicts which are relevant to the child.*
>
> *(2) States Parties shall take all feasible measures to ensure that persons who have not attained the age of 15 years do not take a direct part in hostilities.*
>
> *(3) States Parties shall refrain from recruiting any person who has not attained the age of 15 years into their armed forces. In recruiting among those persons who have attained the age of 15 years but who have not attained the age of 18 years, States Parties shall endeavour to give priority to the oldest.*

Paragraph (1) derives from Article 1 common to the four Geneva Conventions of 1949 and paragraphs (2) and (3) from Article 77(2) of

[392] See for example, M. T. Dutli, "Captured Child Combatants", International Review of the Red Cross, Sept-Oct 1990, pp 421-434, p 424

[393] The exceptions are Oman, Somalia, the United Arab Emirates and the USA

Additional Protocol I to those Conventions. Therefore, the existing jurisprudence and interpretation of the Geneva Conventions and Protocols apply to Article 38, where relevant, and are used in the following analysis.

Article 38 was an innovation in explicitly incorporating international humanitarian law into international human rights law. Previously, the links could only be made by drawing on general provisions that required states parties to human rights treaties to abide by their other obligations under international law.[394]

Article 38(1) applies the dual standard of requiring the states parties both to respect and to ensure respect of these provisions. It therefore establishes positive as well as negative duties both with regard to their own conduct and also in relation to the conduct of others (whether other governments or armed opposition groups). The limitation to "conflicts relevant to the child" is superfluous. It is hard to conceptualise an armed conflict which is not "relevant to the child", even apart from the difficulty of defining "relevant". Even leaving aside the situations in which children are directly affected by attacks, recruitment and so on, armed conflicts tend to have a severe impact on the population in general, on the provision of services, including food, health and education, and on infrastructure, and also tend to affect adults who are relevant to the child, such as fathers and brothers.

The logical interpretation of Article 38(1) is that it simply reinforces the obligations of states to abide by the international humanitarian law by which they are already bound. As such, its value is in making the explicit link between those two branches of international law affecting the human person and in creating the possibility that the Committee on the Rights of the Child, the supervisory body established by the Convention, could examine states' compliance with their obligations under international humanitarian law, at least in so far as relevant to the child. (This is where the limitation does makes sense).

No definition or delimitation of the term "armed conflict" is provided

394 For example, Article 4 of the International Covenant on Civil and Political Rights which permits states to derogate from some of their obligations "In time of public emergency threatening the life of the nation...provided that such measures are not inconsistent with their other obligations under international law..."

in the Convention on the Rights of the Child. It can be argued that a factual definition[395] therefore applies rather than the technical legal definitions of international humanitarian law (see below). In particular, no distinction is made between international and internal armed conflicts. The Convention therefore avoids the problem of governments denying that they are involved in an armed conflict as classified by the 1949 Geneva Conventions or the 1977 Additional Protocols, and therefore denying the applicability of international humanitarian law. One of the consequences of the non-application of international humanitarian law is that armed opposition groups are not bound by any rules since international human rights law is only binding on governments.

Article 38 does not appear to govern recruitment by armed opposition groups since Article 38(3) only addresses states parties recruitment into their own armed forces. However, Article 38(2), particularly when read in conjunction with Article 38(1), does require states parties to the Convention to take all feasible measures to ensure that under-15s do not take a direct part in hostilities, without any apparent limitation to their own armed forces, nor to the type of armed conflict in which the state is engaged (international or internal). The boundaries of what "feasible measures" and "direct" participation constitute are, of course, open to debate. However, this means that a state party to the Convention has an obligation to try to prevent the direct participation of under-15s in hostilities on its own side, in armed opposition groups it is fighting, or on the side of any other government or armed opposition group. In other words, the obligation is not only on the government engaged in an armed conflict but on all those who are parties to the Convention in relation to anyone engaged in an armed conflict.

Article 38(3) is an absolute obligation on the governments not to recruit under-15s into their own armed forces. This should cover any organised armed group which is under the control of the government or which could be under governmental control, and not only the regular armed

[395] That is, by assessing the actual situation on the ground rather than judging whether certain legal conditions are fulfilled or whether the government accepts that it is involved in an armed conflict.

forces. In particular, it should cover any paramilitary forces, local militias, civil defence committees, and so on, under whatever name, which the government encourages or allows: the crucial question is simply whether the group is allowed to carry arms. Any group which meets these criteria can and should have its recruitment regulated by the government in conformity with Article 38.

The requirement in Article 38(3) to "endeavour to give priority to the oldest" when recruiting in the 15–18 age group is a weak obligation. At best, it could be said to require a system which takes account of age in recruiting in this bracket. Its greater use is probably for the purposes of advocacy and moral pressure.

A limitation of Article 38 of the Convention is that it is only binding on States Parties, that is, on governments and not on armed opposition groups. However, the obligation in Article 38(1) incorporates, *inter alia,* Additional Protocol II for those States which are parties to both documents. Furthermore, Article 41 of the Convention provides: "Nothing in the present Convention shall affect any provisions which are more conducive to the realisation of the rights of the child and which may be contained in: (a) The law of a State Party; or (b) International law in force for that State". In other words, if greater protection to children would be given by some other legal instrument to which the state is a party, its provisions will take precedence over those of the Convention on the Rights of the Child. In addition, if the national law sets 18 years as the minimum age for recruitment, the government will be bound by that rather than the lower limit in the Convention.

Since the wording of Article 38 of the Convention on the Rights of the Child was taken from Additional Protocol I, Article 77(2) and that effectively all countries in the world are now parties to the Convention, it is not necessary to examine Protocol I separately as, in effect, this aspect of the Protocol has been superseded by the Convention.

Article 4(3)(c) of Additional Protocol II of 1977 states:

> *Children who have not attained the age of 15 years shall neither be recruited in the armed forces or groups nor allowed to take part in hostilities.*

This provides an unequivocal and total prohibition on recruitment of under-15s into any type of armed forces (governmental or opposition) and any kind of participation by them in hostilities. Thus it does not limit the governmental obligation to taking "feasible" measures to prevent them taking "a direct" part in hostilities, nor does it limit the state's obligation to not recruiting under-15s into its own armed forces only. In addition, the obligation is also imposed on any opposing armed groups.

However, Additional Protocol II only applies to internal armed conflicts in the territory of a state party to Protocol II[396] between governmental armed forces and "dissident armed forces or other organised armed groups which, under responsible command, exercise such control over a part of its territory as to enable them to carry out sustained and concerted military operations and to implement this Protocol" (Article 1(1)). It does not "apply to situations of internal disturbances and tensions, such as riots, isolated and sporadic acts of violence and other acts of a similar nature, as not being armed conflicts" (Article 1(2)). To date, Protocol II has officially been applied only twice – in El Salvador and in the Philippines – because of the high threshold for its application, the reluctance of governments to accept that they are involved in a civil war, and the fact that it only applies to states which are parties to the Protocol.

Finally, in relation to all these provisions it is important to stress that it is not the children who are recruited (even if they volunteer) or who participate in hostilities who are in breach of the provision but the government or other authorities who recruit them or permit their recruitment or participation.

DEVELOPMENTS SINCE THE CONVENTION ON THE RIGHTS OF THE CHILD

At the time of drafting the Convention, the question of the minimum age for recruitment into armed forces and participation in hostilities gave rise to considerable controversy. Many wished to see the minimum age set at 18

[396] See Table 2 for the current list of states parties to Additional Protocol II

years in line with the general age of majority stated in Article 1 of the Convention.[397] In fact, this is the only provision in the Convention which specifies an age lower than 18 years as applicable. Even in this provision, some recognition is given to the 18-year threshold for majority in so far as Article 38(3) requires states "to endeavour to give priority to the oldest" when recruiting in the 15–18 year age range. Since then, a number of steps has been taken to raise the minimum age to 18.

The only regional human rights treaty which addresses the age of recruitment is the 1990 African Charter on the Rights and Welfare of the Child, Article 2 of which defines a child as every human being below 18 years. Article 22 reaffirms the obligation to respect and ensure respect for rules of international humanitarian law, and goes on "(2) States Parties to the present Charter shall take all necessary measures to ensure that no child shall take a direct part in hostilities and refrain in particular, from recruiting any child." Thus governments may not recruit under 18s, and have an absolute duty to ensure that under-18s do not take a direct part in hostilities. Unfortunately, not enough states have yet become parties to this Charter to bring it into effect.

In December 1995, the Council of Delegates of the Red Cross and Red Crescent Movement adopted by consensus a "Plan of Action Concerning Children in Armed Conflict", which includes the Commitment:

> *To promote the principle of non-recruitment and non-participation in armed conflict of children under the age of 18 years.*

This commitment is broken down into three specific objectives:

1. *Promote national and international legal standards (such as an Optional Protocol to the Convention on the Rights of the Child) prohibiting the military recruitment and use in hostilities of persons younger than 18 years of age, and also the recognition and enforcement of such standards by all armed groups (governmental and non-governmental).*

[397] Eight countries have expressed their wish for a higher age by way of declarations on signature or ratification – Argentina, Austria, Colombia, Germany, the Netherlands, Spain, Switzerland and Uruguay.

2. *Prevent children from joining armed forces or groups by offering them alternatives to enlistment.*

3. *Raise awareness in society of the need not to allow children to join armed forces or groups.*

The work involved in preparing and adopting this Plan of Action also led to the adoption (by consensus) of a resolution at the 26th International Conference of the Red Cross and Red Crescent (Geneva, December 1995). The resolution supports the drafting of an optional protocol to the Convention on the Rights of the Child "to increase the protection of children involved in armed conflicts", and:

> *c) also strongly condemns recruitment and conscription of children under the age of 15 years in the armed forces or armed groups, which constitute a violation of international humanitarian law, and demands that those responsible for such acts are brought to justice and punished;*
>
> *d) recommends that parties to conflict refrain from arming children under the age of 18 years and take every feasible step to ensure that children under the age of 18 years do not take part in hostilities;*

Although such resolutions are not legally binding, they are significant because they are adopted at a meeting open to all states parties to the 1949 Geneva Conventions and 1977 Additional Protocols and representatives of all recognised national societies of the Red Cross and Red Crescent Movement.

The optional protocol to the Convention on the Rights of the Child referred to in the resolution and Plan of Action is being drafted by a working group of the UN Commission on Human Rights. The third (and final) session of the working group is due to take place in 1997. The protocol might therefore be adopted by the UN General Assembly in Autumn 1997. States which became parties to the protocol would be bound by its provisions instead of Article 38(2) and (3). To date, a basic agreement seems to have been reached on raising to 18 years the minimum age for participation in hostilities, for compulsory recruitment (conscription) into government armed forces, and for any kind of recruitment into non-governmental (opposition) armed groups. The question of the minimum age (16, 17 or 18 years)

Young revolutionary guard during a demonstration in Iran. Photo: Abbas/Magnum.

for voluntary recruitment into government armed forces is still under discussion. Similarly, whether the prohibition on participation on hostilities should be limited to taking a direct part in hostilities remains unresolved.

WHY 18?

The reason for seeking to establish 18 as the minimum age for recruitment and participation in hostilities is not only based on the general age of majority stated in Article 1 of the Convention on the Rights of the Child. In relation to recruitment, Article 38(3) marks out the 15–18 year age group and exhorts states recruiting in that age bracket to give priority to the eldest. A number of other provisions in international law provide for special protection of those below the age of 18 years, in particular in relation to the non-application of the death penalty to those who were under 18 at the time the offence was committed.[398] In addition, the ILO Convention No. 138 on Minimum Age, 1973, sets 18 years as "the minimum age for admission to employment or work which by its nature or the circumstances in which it is carried out is likely to jeopardise the health, safety or morals of young persons". Although the armed forces are considered to be outside the legal scope of this Convention, the ILO itself has suggested that it "may be applied in corollary to the involvement in armed conflicts".[399]

International humanitarian law provides for the protection of defined groups, such as civilians. Those who may legitimately kill and be killed are the members of armed forces. The recruitment of under-18s makes this group of children lawful objects of attack because they are members of armed forces.

Goodwin-Gill and Cohn support the choice of 18 as the minimum age for recruitment and participation in hostilities on the basis that "participation in the political process is ... a reasonably accurate indicator of the

[398] Convention on the Rights of the Child, Article 37(a), International Covenant on Civil and Political Rights Article 6(5), Additional Protocol I Article 77(5) and Additional Protocol II Article 6(4)

[399] Comments on the Report of the Working Group on a draft optional protocol to the Convention on the Rights of the Child on involvement of children in armed conflicts, contained in UN document E/CN.4/1996/WG.13/2 of 23 November 1995, p 12

moment at which the community as a political body recognises the intellectual maturity of the individual",[400] and shows that the vast majority of countries in all regions of the world have set 18 years as the voting age. They also cite state practice as supporting 18 years as the minimum age for recruitment. Of the 185 states they surveyed, only seven are quoted as having a lower age for compulsory recruitment, and only six a lower voting age.

They argue: "Given the essentially political dimension to armed conflict, whether national or non-international, the choice of 18 as the moment of transition to adulthood does not seem unreasonable. Indeed, in principle, it would seem wrong to condemn the unenfranchised to die as a consequence of political decisions on which they can exercise no influence." Furthermore, they argue that the international human rights and humanitarian law prohibition of the death penalty on those under 18 years at the time of the commission of the offence, applicable alike in times of peace and war, "acknowledges the reduced ability of those under 18 to appreciate the nature of their action in the context of criminal responsibility. The same consideration, however, is not given to the capacity of the child or young person to evaluate the reasons for death in combat."[401]

The physical, emotional and social impact of involvement in armed conflict of those under 18 years is too well documented in other chapters of this book to warrant repetition here. Furthermore, the evidence of this book is that children are recruited predominantly because not enough adult recruits are forthcoming, or in order to use them as spies or to commit atrocities. These are not reasons compatible with the Convention on the Rights of the Child. In addition, the active participation of some children exposes other children, particularly in the conflict zones, to pressure to join one or other side, and to suspicion of involvement making them liable to attack, interrogation and other harassment. In addition to reducing the actual participation of under-18s, reducing the suspicion of such participation should help to protect all children in affected areas from attack, harassment and detention.

[400] G. Goodwin-Gill and I. Cohn: Child Soldiers (Clarendon Press, Oxford, 1994), p 7

[401] Ibid, p 9

CHAPTER 8

Strategies to Prevent Recruitment

EXAMPLES HAVE ALREADY BEEN given of tactics used to avoid forcible recruitment and by families to distance their children and thus dissuade them from volunteering or spontaneously taking part. Essentially, however, such tactics do nothing to tackle the phenomenon of child recruitment, addressing only the question of who is affected, with, as discussed in Chapter 3, certain systematic consequences. There are, however, instances where action has been taken which has had a more general effect in discouraging child recruitment, and from them may also be drawn some indication of further approaches which might prove fruitful. These are considered in this section, by reference to the main actors whose assistance may be enlisted. However, the sections are not self-contained and, in particular, the role of different actors, including those of international organisations and specialised agencies, are woven into the sections rather than being treated under a separate heading.

ARMED FORCES AND ARMED OPPOSITION GROUPS

Since most of the recruitment is carried out directly or indirectly by the armed forces and armed opposition groups themselves, without their co-operation it will not be possible to eliminate child recruitment. Obviously those who deliberately choose to make use of child soldiers are not in the short term going to be amenable to persuasion, but the evidence does not suggest that such cases are in the majority. More often, the recruitment of children seems to be the consequence of negligence and inadequate procedures rather than of conscious desire. Where governments, and armed opposition groups, have proper recruitment procedures, children are less likely to become involved.

As shown above, there is a body of informed military opinion which holds that young recruits are at best a mixed blessing and at worst a positive liability. In purely military terms there is no clear evidence of any particular effectiveness of child soldiers in general; quite a lot suggests the contrary. If this message could be spread more widely, all sorts of armed forces might be encouraged to consider improvements to their recruiting methods, not just to avoid international opprobium, but with the more direct result of improving their own efficiency by eliminating the enlistment of unsuitable recruits.

Further, enlightened self-interest points towards the non-use of children by all parties in view of the emerging evidence that the widespread involvement of child combatants is in itself a factor in the continuation of armed conflicts, in addition to their propensity to commit atrocities and the links to violent criminal activity. The mutuality of interests of the opposing forces in this regard reinforces the idea that local agreements on a minimum age of recruitment are a valid option, and one which communities, other countries, international and national organisations and non-governmental organisations should encourage. Further, such agreements, if implemented, would have a beneficial effect for all children, particularly those in or from the conflict zones, by reducing the direct pressures on them and by reducing the general level of their being suspected and, therefore, of their harassment and other ill-treatment. Communities have a direct and particular interest in encouraging such agreements.

GOVERNMENTS

The self-evident action which governments can take is to outlaw the recruitment of children in all government armed forces, including militias/civil defence forces, and to introduce effective recruitment procedures. In practice, this cannot be effective unless accompanied by a ban on all forms of forced recruitment. Many of the practices described in this report amount to violations of fundamental human rights, arbitrary detention, disappearance, torture and summary executions, which governments are under a duty to prevent and remedy. It is, however, easy to imagine that the promulgation of legislation will not affect the situation on the ground unless accompanied by legal remedies and institutions strong enough and willing to tackle abuses.

> *... on the subject of this illegal procedure, 596 reports for May and June 1995 have been received about youths who have been recruited by force. As a result of intervention and pressure from this institution, 333 people have managed to be set free, 148 of these being under 18. All this in spite of the ineffective intervention of some judges ...*[402]

Apart from the question of recruitment methods themselves, governments can play a major part in preventing recruitment into both their own armed forces and armed opposition groups. They can seek to insulate the civilian population from the physical and economic consequences of the conflict. Keeping schools open, or establishing, renovating and reopening schools,[403] not only gives the young a civilian focus but avoids the development of the situation whereby the gap is filled by institutions linked to armed opposition groups. The availability of schooling for refugee and displaced children also needs to be addressed. Other governments, international and national organisations and non-governmental organisations can help both in terms of advocacy and of provision.

Recognition of conscientious objection to military service and provision for alternative civilian service both help to counter the overwhelming militarisation of society by validating alternatives. The establishment of the necessary procedures to enable this, will also assist in the general regularisation of recruitment procedures and the requirement to ensure that potential recruits know their rights and how to claim them.

Again, quite apart from preventing it from taking place, under-age recruitment is hard to detect and monitor where individuals lack documen-

[402] Report of the Human Rights Ombudsman quoted in the case study for Guatemala (translated from the Spanish)

[403] Case studies for Lebanon and Liberia

tary proof of age. Article 7(1) of the Convention on the Rights of the Child, which has achieved near universal ratification, requires states parties to provide proper birth records. Governments could do much to deter underage recruitment by thorough and universal implementation of this obligation, which should apply also to refugees and internally displaced persons. Donor governments, international and national organisations and non-governmental organisations could play a major part in encouraging and assisting in the establishment of systematic procedures for universal, routine registration of births. However, even if such procedures were implemented immediately, registration of existing children also needs to be addressed.

Governments can also play a part in putting pressure on armed opposition groups in relation to their conduct by becoming parties to relevant international humanitarian law conventions and applying them in internal armed conflicts, thus making such groups also accountable under humanitarian law. Other governments can also bring pressure to bear on such groups, and those which are party to the Convention on the Rights of the Child and relevant international humanitarian law treaties have an obligation to do so.

At an even more fundamental level, as long as armed conflicts occur (particularly internal ones), children are likely to become involved in them; the longer they continue, the greater and the younger the child participation tends to become. These may not be the best arguments for governments to address the causes of conflicts and to bring them to an early end. However, the point has to be made that all the other strategies combined will not by themselves solve the problem of child soldiers.

One obvious step would seem to be that it is in the self-interest of governments to regulate the flow of lightweight automatic weapons and other weapons which are light and simple enough for children to use to such devastating effect, whether in armed conflicts or in violent crime.

COMMUNITIES, NATIONAL AND INTERNATIONAL NON-GOVERNMENTAL ORGANISATIONS

Sometimes the attitude of the local community may be enough to dissuade illegal forced recruitment. One case study reports a claim:

that the Army has only changed its recruitment methods in some places where the inhabitants have opposed it ... However, in areas where they have not met any opposition they continue to use forced recruitment.[404]

Another reports "the notable decline in the number of 'levies' [army recruiting drives] in the areas under the jurisdiction of the parish churches which denounced this activity."[405]

A few individual cases of forced recruitment of minors have been remedied through *habeas corpus* proceedings brought by the national Committee for the Defence of Human Rights, or by families with its support, but such proceedings cannot effect structural change.[406]

The populations affected – ethnic groups, or mothers of actual or potential recruits – have in some places formed organisations which have tried to bring political pressure to bear against governmental recruitment of children by writing letters, organising demonstrations, appealing to parliament and the army leadership, public denunciation of forced recruitment, receiving testimonies of mothers whose sons have been taken, and even going on hunger strike.[407] If the children can prove that they are under age according to the national law, they are quite often returned to their communities, especially if their parents or guardians claim them and the human rights non-governmental organisations protest applying a great deal of pressure on the army.[408]

Outside organisations, too, can wield influence. One instance is recorded where protests from aid organisations led to the return of boys and men forcibly recruited from a refugee camp,[409] and there are reports elsewhere of the negotiation of agreements on non-recruitment of under-15s between humanitarian organisations and local commanders.[410] Such organisations

[404] Case study for Guatemala (translated from the Spanish)
[405] Case study for Peru (translated from the Spanish)
[406] Case study for Honduras
[407] Case study for El Salvador, Guatemala and Paraguay
[408] Case study for Guatemala
[409] Case study for Burma/Myanmar
[410] SPLM/Operation Lifeline Sudan Agreement on Ground Rules, July 1995

can also help by informing children of their right not to be recruited and helping to counter some of the pressures to join the armed struggle.[411] Much more, however, could be done by international non-governmental organisations (in particular, children's and humanitarian ones in the field) and by UNICEF and UNHCR through monitoring and reporting on abuses, raising the issues with those in authority, supporting local groups working on the issues, and educating children and young people, and the military and other authorities.

By contrast to all this, the greatest succour which can be given to systematic forced and child recruitment is the silence of popular opinion and international organisations, however good the excuses for such silence:

> *The most active Human Rights Organisations ... both admitted having helped mothers to demand their forcibly recruited sons, but neither had done anything to discourage or prevent minors of age from recruiting voluntarily. Their representatives admitted that in this respect lots of violations of Rights of Children were committed, but said apologetically that they had so much work to do on those days, so many more urgent challenges ... most [citizens] seem to have felt that fighting the war was not so very different from the other duties and jobs children traditionally perform among the poor sectors of society. Children endure all kinds of hardships, and are deprived in so many ways that their participation in the armed forces did not attract any major attention by the society and its institutions. Anyhow, the great majority of children on both sides of the conflict were children that never really "counted/mattered" in the eyes of society ... a concept prevails that in the war situation the constitution and international conventions lose their validity.*[412]

FAMILIES

The maintenance of families and the attitude of those families are one of

[411] Human Rights Watch/Africa, Children's Rights Project: Children of Sudan (September 1995), p 85

[412] Case study for El Salvador

A boy playing with a toy gun in South Africa 1994. Counteracting the glorification of war and violence is one way of preventing children's participation in armed conflicts. Photo: Gamma/IBL.

the key factors in reducing child recruitment. Those parents who see the military struggle as "barbaric and uncivilised", or are opposed because of their religious background, or "because of their experiences about the devastating effect of war",[413] will do everything they can to ensure the non-involvement of their children. In some cases this may include assigning sons to the monkhood for varying lengths of time in order to protect them from recruitment.[414] Similarly protected are those who

> *come from families where the family relationships are stable and where parents have positive and healthy attitudes toward life itself; where, for the parents, the worth of their children's lives is above their own national, patriotic or political interests.*[415]

It might seem that the families already do what they can, particularly in the matter of adopting evasive strategies. Many children, however, fail to benefit from such protection simply because they have no family background. The importance of supporting and maintaining family structures cannot, therefore, be overemphasised: not only physically, but also economically, socially and emotionally. Family reunification programmes for street/orphaned and abandoned children[416] can reduce the pool of ready recruits. Governments, agencies and non-governmental organisations can all play a part in such endeavours.

Where children and their families know the national and international law concerning the age of recruitment and participation in hostilities, children are less likely to be recruited in the first place. It may also enable them to challenge illegal recruitment, particularly where there are governmental institutions able and willing to bring legal cases.[417] Banding together to form mutual support groups or campaigning organisations increases the likelihood of success and of effecting changes in policies and practices.

[413] Case study for Liberia
[414] Case studies for Burma/Myanmar and Cambodia
[415] Case study for the former Yugoslavia
[416] Case study for Liberia
[417] Case study for Guatemala and Honduras

Ignorance of the national and international laws means that this factor is not even considered by the children themselves or by their families.[418] Informing children of their rights, including non-recruitment of anyone under 15 years by any party, may enable unaccompanied children to withstand pressure to join. Too often even if they know the law, families do not believe in its efficacy in a conflict situation (with justification) and, if children (or these particular children) are not valued by society, their involvement will be harder to prevent.[419]

It is clear that the attitude of families and communities is a crucial factor both in the immediate situations and the longer term. As long as the involvement of children is accepted as inevitable, the situation is unlikely to change fundamentally. On the other hand, where communities oppose such involvement and they have a modicum of power or influence, the incidence of such involvement is negligible. The fact that the overwhelming picture is of the poorest, least educated and most disadvantaged children being the recruits, undermines any argument of a lower age for recruitment or participation being culturally acceptable, as does the existence of national legislation which does not take account of such supposed cultural variations. The preventive strategies employed by families reinforce the distinction between functional families with a measure of education, income and influence who can act to protect their own children from compulsory recruitment and volunteering, and dysfunctional, illiterate and poor families, or children without families, who are unable to exercise such choices and strategies. Not only are they, therefore, more at risk initially, but the absence of children from the first group leaves them under greater pressure. The need for systematic protection for all children cannot, therefore, be over-emphasised.

[418] Case study for the Philippines

[419] Case study for El Salvador

CHAPTER 9

Recommendations

LEGAL ISSUES

1. International law should recognise 18 years as the minimum age for recruitment (compulsory or voluntary) into any kind of armed forces and armed groups and for any kind of participation in hostilities. The proposed Optional Protocol to the Convention on the Rights of the Child on involvement of children in armed conflict should establish these standards and be adopted as soon as possible.
2. Where a Government derogates under a human rights treaty, it should not be able to deny the applicability of at least common Article 3 of the Geneva Conventions of 1949.
3. Governments should become parties to Additional Protocol II to the Geneva Conventions of 1949, and governments and opposition groups should be encouraged to apply the Protocol.
4. African states should become parties to the 1990 African Charter on the Rights and Welfare of the Child, which sets 18 years as the minimum age for recruitment and participation in hostilities.

THE REGULATION OF RECRUITMENT

5. All governments and armed opposition groups who currently have persons under 18 years in their armed forces should be urged to demobilise them immediately, and to refrain from recruiting under-18s in future. In current conflict/internal disturbance situations, all parties should be urged to make local agreements to this effect. The mutual nature of such agreements should encourage parties to keep to them and, in addition to reducing the actual participation of

under-18s, should help to protect all children in the affected areas from harassment and detention (by reducing the suspicion of participation). All governments, as well as relevant international and regional organisations, specialised agencies and international and national non-governmental organisations, can assist in advocating for and negotiating such agreements.

6. Governments should introduce proper recruitment procedures for their armed forces:

   (a) Where conscription/compulsory recruitment exists, it must be given a legal basis, with a clear minimum age, which should be not less than 18 years. The law should be publicised both to those recruiting and to the public at large in ways which will ensure that those liable to conscription, and those not liable to it, know their rights and duties. Proper safeguards, including requirement of proof of age, and military disciplinary or criminal penalties for infringements, should be available and enforced to ensure that under-age recruitment does not take place.

   (b) The role of local militias/civil defence forces as recruiting agents should be eliminated. Their recruitment ages and procedures should be brought into line with those for regular government armed forces, and the provisions of UN Commission on Human Rights resolution 1994/67 implemented. The resolution recommends: "(a) Civil defence forces shall only be deployed for the purpose of self-defence; (b) Recruitment into them shall be voluntary and shall be effectively controlled by public authorities; (c) Public authorities shall supervise their training, arming, discipline and operations; (d) Commanders shall have clear responsibility for their activities; (e) Civil defence forces and their commanders shall be clearly accountable for their activities; (f) Offences involving human rights violations by such forces shall be subject to the jurisdiction of the civilian courts". Effective, practical measures to ensure the implementation of these recommendations require further consideration.

   (c) Where there is conscription/compulsory recruitment, provision should be made for exemption from military service on grounds of conscience and for an alternative, non-military, non-punitive, service for conscientious objectors to military service which is compatible with the reasons for the objection, in accordance with UN Commission on Human Rights resolution 1995/83.

(d) Voluntary recruitment should require appropriate safeguards both in relation to age and to voluntariness with the obligation on the recruiter to obtain confirmation of the full and informed consent of the individual and of any persons (such as parents) with legal responsibility for the individual volunteering.

OTHER PREVENTIVE MEASURES

7. The routine provision of official birth records is an essential prerequisite for the prevention of under-age recruitment into government armed forces (whether compulsory or voluntary). It is also required by Article 7(1) of the Convention on the Rights of the Child.

(a) Governments should, as a matter of priority, establish effective and universal systems for registering births and providing the necessary documentation.

(b) UNICEF, other agencies (including UNHCR in relation to persons of concern to them) and non-governmental organisations should assist governments in establishing such systems.

(c) In the meantime, and for those born before such systems are fully operational, UNICEF, UNHCR, other agencies and non-governmental organisations should seek to assist individuals/families in establishing ages and ensuring the provision of appropriate documentation.

8. UNICEF, UNHCR, other agencies and non-governmental organisations should inform themselves about the age(s) for compulsory and/or voluntary recruitment under the relevant national law, and ensure that:

(a) it is in conformity with international law;

(b) persons with whom they are in contact know their rights –

including the minimum age(s) relevant, and exempted categories if any – and what recourse is available to them if they are unlawfully recruited;

(c) the law is applied in practice and, if not, should take this up with the appropriate national authorities, the Committee on the Rights of the Child and other UN mechanisms such as the UN Commission on Human Rights Special Rapporteurs and Working Groups on Summary Executions, Arbitrary Detention, Torture, Disappearances, Sale of Children, and relevant Country Rapporteurs.

9. The most effective ways for Governments to reduce volunteering into opposition armed groups are:

(a) Not to attack or subject to violent harassment, children themselves, their families and homes.

(b) Where children are involved in armed opposition groups, to treat them in accordance with recognised international standards on capture, and also with the Convention on the Rights of the Child, and allow regular access to all child detainees by the International Committee of the Red Cross.

(c) To provide access to education and vocational training for all children.

(d) To address the economic, social and political causes of the conflict.

10. UNHCR and other agencies and non-governmental organisations should make provision, in particular in the form of education and vocational training, for children and young people, including refugee and internally displaced children in order to reduce volunteering .

11. The Committee on the Rights of the Child and the thematic and country-specific mechanisms of the UN Commission on Human Rights, should specifically monitor, respond to and report on violations of the rights of children who are soldiers.

ISSUES REQUIRING FURTHER CONSIDERATION

12. Consideration should be given to the need for and advisability of a specific UN mechanism, such as a Special Rapporteur of the UN Commission on Human Rights, to address the issue of child soldiers.

13. The UN Sub-Commission on Prevention of Discrimination and Protection of Minorities should undertake a study of the treatment of recruits in government armed forces with a view to examining its compatibility with human rights standards, as well as the consequences of inhuman and degrading treatment for the recruits and for the subsequent behaviour of the armed forces.

14. The Committee on the Rights of the Child should consider the questions of applicability to children of:

    (a) emergency and anti-terrorism legislation;

    (b) military law, punishment and discipline (including in military schools), and should also consider the question of the age of criminal responsibility, in particular in relation to situations of armed conflict, states of emergency and terrorism, and their compatibility with the Convention on the Rights of the Child. Governments should review their legislation to ensure its conformity with the Convention and other international standards.

15. Governments should regulate the flow of lightweight automatic weapons, and other small arms.

DEMOBILISATION, REHABILITATION AND SOCIAL REINTEGRATION

16. At the time of demobilisation, and to ensure the rehabilitation and reintegration of child combatants into civil society, the provisions of the Convention on the Rights of the Child should constitute the guiding principles in all actions to protect their rights and welfare.

(a) In particular, the international community must advocate governments and other concerned groups to ensure that the participation of children in conflict is recognised in peace agreements and related documents, in order that their needs are not neglected and are incorporated as a matter of principle in plans for demobilisation, rehabilitation and social reintegration.

(b) Advocacy and action on behalf of child soldiers should be implemented with due regard for the children's right to privacy, should respect their dignity, protect their physical and emotional well-being, and ensure that no children are exploited with the purpose of drawing attention to any particular situation/programme. Agencies involved with the welfare of child soldiers should establish guidelines and procedures that reflect professional ethical standards to address these issues.

(c) The children should be informed of solutions that are being considered to address their circumstances and, with due regard for their age and stage of maturity, their participation should be incorporated into defining strategies for their rehabilitation and social reintegration.

17. When it is known that children are to be demobilised, international organisations, in particular UNICEF and UNHCR, and international and local non-governmental organisations should implement programmes of family tracing and reunification as a matter of urgency, to ensure the return of children to their communities of origin.

(a) Alternative systems of care that are appropriate to the culture and circumstances of the children should be established to meet the needs of children who, for whatever reason, cannot be reunited with their families. The institutionalisation of children should be discouraged, and only implemented when all alternative forms of care have been considered.

(b) All such programmes should be implemented in partnership with the appropriate governmental authorities to ensure that there is a capacity to meet the needs of the children, and that their well-being is monitored.

18. The international community, in particular UNICEF and the non-governmental organisations in partnership with local organisations, should ensure that communities are not left alone to bear the burden of rehabilitation and social reintegration of the children, and should recognise the crucial role they can play in this process.

(a) Programmes that incorporate the participation of the children, their families and communities, and enable them to normalise their daily lives and prevent further recruitment of under-age combatants should be implemented. In this regard, projects directed at implementing sustainable initiatives that provide support to the family and community and improve the economic and social conditions that influenced the initial recruitment of the children should be considered a priority.

(b) Programmes of education and vocational training should also be a priority concern in the rehabilitation and social reintegration of child soldiers. Alternatives to traditional educational systems will need to be considered, and additional training for teachers made available. Opportunities for education and vocational training must be appropriate to the circumstances of the communities, and not lead to unrealistic aspirations on the part of the children. For older children, or those who due to long periods of time spent with the military have difficulty in returning to school, the creation of opportunities for self-employment to compensate for their lack of skills should be considered.

(c) The needs of child soldiers should be addressed within overall planning and service provision for all children affected by armed conflict, to ensure that they are not marginalised within their communities. However, special programmes to assist the social reintegration of especially vulnerable groups (e.g. girl soldiers; disabled child combatants) may be required in some circumstances. Appropriate methods of coping with these situations should be determined through dialogue with the children, their families and communities.

19. Governments should ensure that follow-up of demobilised child soldiers takes note of community attitudes towards the culpability of children who were actively involved in conflict, and ensure that they are awarded due protection within the national juvenile justice system, and in accordance with the provisions of the Convention on the Rights of the Child. International and local non-governmental organisations should assist in these procedures, and intervene where necessary to protect the rights and well-being of the children, and to promote reconciliation and understanding of the causes and consequences of children's participation in conflict.

CHAPTER 10

Executive Summary

THE APPLICABILITY OF THE UN CONVENTION ON THE RIGHTS OF THE CHILD

This book focuses on the causes and consequences of the involvement of children in armed forces and armed opposition groups and as combatants in armed conflict. The principles of action to effect change for children who have been recruited or have participated in armed conflict and to promote their best interests must be guided by their needs and those of their families and communities as described in the Convention on the Rights of the Child. The provisions of the Convention, to which nearly every state is now a party, should constitute the guiding principles in all actions to protect the rights and welfare of child recruits and combatants. Where children are involved in government armed forces or armed opposition groups they should be treated in accordance with recognised international standards on capture, and also in accordance with the provisions of the Convention. The Convention should be the basis to ensure that the participation of children in conflict is recognised in peace agreements and related documents, and that their needs are incorporated in plans for demobilisation, rehabilitation and social reintegration. (Chapters 4, 5)

THE CONSEQUENCES OF PARTICIPATION IN CONFLICT FOR CHILDREN

The principal consequence for the vast majority of children who participate in armed conflict is separation from their families and communities. This separation occurs in violent circumstances at a period in their lives when children have most need of the care and support that family life provides. Both government forces and armed opposition groups should not attack or subject to violent harassment children themselves, their families and homes. (Chapters 2, 3, 5)

As members of both government forces and armed opposition groups, children's deaths are the result of active involvement in combat where their inexperience and, frequently, lack of training results in high casualty rates. They are commonly used on the front lines, and their size and agility lead them to be given particularly hazardous assignments. Suicide missions are considered the particular province of adolescents. Many children who are injured in combat are left to die from their wounds, or are shot. Those who are too weak to keep up with the group, who attempt to escape, either to avoid recruitment or to desert, are usually executed. Children die from injuries incurred during beatings, whether as a punishment or to break the spirit of new recruits. They are also more prone to die from starvation and preventable diseases contracted in the unhygienic conditions in which they live. (Chapters 2, 4)

With the exception of certain armed forces and armed opposition groups whose treatment of children is more humane, the treatment and training of child soldiers involves a high degree of risk for their physical well-being, especially for the youngest amongst them. Their bodies are still developing and they are thus at greater risk of injury and disability from the privations that are common in military life. These include poor diet, insanitary conditions and inadequate health care, and the rigours of harsh training routines and excessive punishments that can leave them weakened and debilitated. (Chapter 4)

There are numerous instances of children being routinely administered drugs and alcohol, particularly before a battle. Sexual abuse of both boys and girls is not infrequently reported, entailing for many the risk of sexually transmitted diseases, HIV/AIDS, and pregnancy for the girls. Where the pregnancy is terminated in insanitary conditions, this can put the girls at additional risk of associated disorders. (Chapter 4)

The most frequent injuries suffered by child soldiers are loss of limbs, loss of hearing and blindness. These disabilities will impose additional hardships in the future, compromising their chances to take advantage of educational and vocational programmes, and impeding their social reintegration as they become an additional burden for an already impoverished family. (Chapters 4, 5)

Over and above the specifically physical consequences the children suffer, they are also subjected to degrading and humiliating treatment to subordinate them to authority. Such treatment can influence their psychological well-being, destroying their self-esteem and leading to violent solutions to problems. Some children commit suicide following such events. (Chapter 4)

Given these consequences of recruitment and participation in conflict and the fact that the recruitment of children into armed opposition groups often fuels under-age recruitment into government armed forces and *vice versa*, the recruitment of children should stop. In addition, the active participation of some children exposes other children, particularly in the conflict zones, to pressure to join one or other side, and to suspicion of involvement making them vulnerable to attack, interrogation and other harassment. All governments and armed opposition groups who currently have persons under 18 years in their armed forces should be urged to demobilise them immediately, and to refrain from recruitment of under-18s in future. In current conflict/internal disturbance situations, all parties should be urged to make local agreements to this effect. The mutual nature of such agreements should encourage parties to keep to them and, in addition to reducing the actual participation of under-18s, by reducing the suspicion of such participation, should help to protect all children in the affected area from attack, harassment and detention. All governments, as well as relevant international and regional organisations, specialised agencies and international and national non-governmental organisations can assist in advocating for and negotiating such agreements. (Chapters 4, 8)

DEMOBILISATION, REHABILITATION AND SOCIAL REINTEGRATION

The demobilisation of child soldiers is seen as essentially the preliminary phase in a process of rehabilitation and social reintegration. Such a process cannot be accomplished without recognition of the participation of children in conflict. This is a requirement of local communities, government and armed opposition groups involved in the conflict, and the international community. It is only when children are demobilised that the reality of their experiences in conflict and the consequences for their well-being can be recognised and addressed. (Chapter 5)

The effects of participation in conflict cannot be explained in straightforward terms. They are principally described as child welfare concerns that have an impact on the children's immediate social and cultural environment, influence both physical and emotional well-being, and limit their opportunities for education, vocational/skills training and, ultimately, their chances for employment. In almost all instances their experiences as child soldiers are described as having a negative effect on these varied aspects of their lives that would normally contribute to their healthy development and as compromising their capacity to reintegrate into their families and communities. (Chapter 5)

The family is considered the children's primary resource, and therefore family tracing and reunification, and the association of children to their communities of origin, should be implemented as a matter of urgency when it is known that children are to be demobilised. (Chapter 5)

Re-attachment to family and community will determine to a large extent effective social reintegration. As a result of the conflict, however, the majority of families are impoverished. Thus programmes addressing rehabilitation and social reintegration should encourage the participation of families and communities. They should also incorporate support to their families and communities, not only to the children. These should be directed at implementing sustainable initiatives that improve the economic and social conditions that in most cases influenced the initial recruitment of the children. (Chapter 5)

Resources available to meet children's needs, both during and after conflict, are limited and there is little likelihood of resources being directed specifically at child soldiers. In any event, all children in the affected area have suffered deprivations due to the war. Thus a more effective and equitable approach will be to address the needs of former child soldiers within overall planning and service provision for all children affected by armed conflict. This would reduce any risk of marginalising children who have been directly involved in combat. A similar consideration will apply for those children who, for whatever reason, cannot be reunited with their families. For them, alternative systems of care must then be considered that meet the children's needs, which are culturally acceptable and, again, avoid any possibility that they will be marginalised. The institutionalisation of children should be discouraged, and only implemented when all alternative forms of care have been considered. (Chapter 5)

However, some children will be especially vulnerable, due to particular experiences and events. Most evident will be children whose disabilities pose a financial burden for families already experiencing economic hardship. Also, children who have been seriously affected by their experience of traumatic events may require special assistance. Another group for whom social reintegration may prove difficult is girl soldiers with children. Whilst special programmes to assist the social reintegration of vulnerable groups should be considered, dialogue with their families and communities should determine appropriate methods of coping with these situations, in particular addressing the consequences of "traumatic events" and other conflict-related problems. (Chapter 5)

The case studies made repeated reference to the linkage between education, employment opportunities, and the economic security of the children's families in determining successful social reintegration, and preventing re-recruitment. Without education, children's future prospects for employment are limited, and the army may be their only opportunity to earn money and contribute to the family economy. Education is also the system within which children's lives can be normalised. Through meaningful and productive activity they can be helped to overcome their experiences and develop an identity separate from that of the soldier. Programmes of education and vocational training should, therefore, be seen as a priority concern in the rehabilitation and social reintegration of child soldiers. Given the potential of large numbers of demobilised child soldiers in some situations, their varying ages, experiences and their own attitudes towards education, alternatives to traditional educational systems will need to be considered, and additional training for teachers made available.

Opportunities for education and vocational training must be appropriate to the circumstances of the communities, and not lead to unrealistic aspirations on the part of the children. For older children or those who, due to long periods of time spent with the military, have difficulty in returning to school, the creation of opportunities for self-employment to compensate for their lack of skills should be considered. (Chapter 5)

The children should be informed of solutions that are being considered to address their circumstances and, with due regard for their age and stage of maturity, their participation should be incorporated into defining strategies for their rehabilitation and social reintegration – see Article 12 of the Convention on the Rights of the Child. (Chapter 5)

Communities cannot be left alone to bear the burden of rehabilitation and social reintegration. Of principal concern is that there are implemented programmes of assistance which have long-term sustainability for these most vulnerable victims of armed conflict, and for their families and communities who have suffered extreme deprivation as a result of the participation of their children in the conflict. Organisations directly involved with the welfare of children affected by participation in armed conflict – principally UNICEF, but also UNHCR and international and local non-governmental organisations – should identify procedures to address the worst effects of participation in armed conflict, and move towards involving the children, their families and communities in processes to effect peaceful strategies. (Chapter 5)

PREVENTING RECRUITMENT OF CHILDREN

The main reason why children are recruited into government armed forces is not because children as such are wanted but because of shortage of manpower and inadequate recruitment procedures. Recruitment of children particularly occurs when forced recruitment, such as press-ganging, is practised, when large numbers of soldiers are required and when the conflict is prolonged. Governments can prevent or minimise under-age recruitment by introducing proper recruitment procedures and prohibiting forced recruitment. Where conscription/compulsory recruitment occurs, it must be given a legal basis, with a clear minimum age, which should be not less than 18 years. The law should be publicised both to those recruiting and to the public at large in ways which will ensure that those liable to conscription, and those not liable to it, know their rights and duties. Proper safeguards, including requirement of proof of age, and military disciplinary or criminal penalties for infringements, should be available and enforced to ensure that under-age recruitment does not take place. (Chapter 2)

To require proof of age presupposes that individuals have a birth registration or identity document which can be used to demonstrate their age. Without such a document, genuine problems about under-age recruitment arise, but the non-availability of proof also encourages unscrupulous disregard of age requirements. The routine provision of official birth records is required by Article 7 of the Convention on the Rights of the Child. It is an essential prerequisite for the prevention of under-age recruitment (whether compulsory or voluntary) into government armed forces. Governments should, as a matter of priority, establish effective and universal systems for registering births and providing the necessary documentation. UNICEF, other agencies (including UNHCR in relation to persons of concern to them), national and international non-governmental organisations and donor governments, can assist in establishing such systems. In the meantime, and for those born before such systems are fully operational, governments, UNICEF, UNHCR, other agencies and non-governmental organisations should seek to assist individuals/families in establishing ages and ensuring the provision of appropriate documentation. (Chapter 2)

Where there is provision for both conscription and voluntary recruitment with a lower entry age for volunteers, under-age conscripts can be passed off as volunteers or pressured into volunteering. Not only should clear minimum ages be established, but the age for volunteers should be the same as for conscripts. In any case, procedures are required to ensure that volunteers are just that. Governments should establish appropriate safeguards for voluntary recruitment both in relation to age and to voluntariness with the obligation on the recruiter to obtain confirmation of the free, full and informed consent of the individual and of any persons (such as parents) with legal responsibility for the individual volunteering. (Chapter 2)

In some countries recruitment for government armed forces is carried out indirectly by local militias/civil defence forces or others. Such indirect systems of recruitment are prone to abuse, particularly when combined with the requirement to produce a specified number of recruits, rather than named individuals. The role of local militias/civil defence forces as recruiting agents for government armed forces should be eliminated. The recruitment ages and procedures for all armed militias/civil defence forces themselves should be aligned with those for regular government armed forces and with UN Commission on Human Rights resolution 1994/67, including the requirement that all recruitment into them be voluntary. (Chapter 2)

There is a need for children themselves and for their families to be informed about their rights and how to enforce them, and for government recruitment practices to be monitored by national and international bodies, to ensure conformity with international law in both law and practice. A lack of national and international response has permitted widespread child recruitment. Governments should include specific information on their recruitment laws and practices in their reports to the Committee on the Rights of the Child. UNICEF, UNHCR, other agencies and national and international non-governmental organisations should inform themselves about the age(s) for compulsory and/or voluntary recruitment under the relevant national law, and ensure: (a) that it is in conformity with international law; (b) that persons with whom they are in contact know their rights – including the minimum age(s) relevant, and exempted categories if any – and what recourse is available to them if they are unlawfully recruited; (c) that the law is applied in practice and, if not, should take this up with the appropriate national authorities, the Committee on the Rights of the Child and other UN mechanisms such as the UN Commission on Human Rights Special Rapporteurs and Working Groups on Summary Executions, Arbitrary Detention, Torture, Disappearances, Sale of Children, and relevant Country Rapporteurs. (Chapter 8)

The lack of an international mechanism systematically monitoring, taking up and reporting on the recruitment of children, their treatment in armed forces and on capture, could be at least partially addressed by the use of existing human rights mechanisms. The Committee on the Rights of the Child, thematic and country-specific mechanisms of the UN Commission on Human Rights, should specifically monitor, respond to and report on violations of the rights of children who are soldiers.

Consideration should be given as to the need for and advisability of a specific mechanism to address the issue of child soldiers. (Chapter 8)

Families can help to counter societal and peer group pressure on children to join armed forces or armed opposition groups. Society itself can also help by reducing the militarisation of daily life and providing alternative models to the glorification of war and military leaders. Specifically, provision should be made for exemption from military service on grounds of conscience and an alternative, non-military and non-punitive form of service for conscientious objectors to military service, in accordance with UN Commission on Human Rights resolution 1995/83. (Chapters 2, 8)

Children become soldiers primarily because they are available, have less power than adults to oppose recruitment and are, therefore, more likely to be forced, intimidated or persuaded into joining government armed forces or armed opposition groups. Child recruitment falls unequally. Those most likely to be recruited, whether willingly or unwillingly, are: (a) first and foremost, children separated in any way from their families; (b) the economically and socially deprived (the poor, whether rural or urban, those without access to education, vocational training, and a reasonable standard of living); (c) other marginalised and disenfranchised groups (such as street children, certain minorities, refugees and the internally displaced); and (d) children from the conflict zones themselves. The factors leading to child recruitment usually occur in combination. Governments, UNICEF, UNHCR, other agencies and organisations, national and international non-governmental organisations, separately and together, all have roles to play in maintaining or reuniting families, providing access to education and vocational training for all children, ensuring adequate economic provision so that such access is real rather than theoretical and that children have the prospect of a livelihood (Convention on the Rights of the Child Articles 27 and 28). The particular needs and vulnerability of separated children, refugees and internally displaced children and families, need to be addressed. (Chapters 2, 3)

Children and their families need to know their rights, including the minimum ages for legal recruitment into government armed forces and that no-one under the age of 15 years should be in any armed force or armed group or participating in hostilities, and to whom they can complain or appeal (Convention on the Rights of the Child, Article 38). Governments, UNICEF, UNHCR (in relation to persons of concern to them), national and international non-governmental organisations should assist in the protection of all children by ensuring that they know their rights in this regard and the means of redress open to them. (Chapter 8)

The single major factor for children volunteering into armed opposition groups is their personal experience of attack, ill-treatment or harassment of themselves or their families by government armed forces. Sometimes they join armed opposition groups in search of revenge but more often it is a desire for protection because of their sense of vulnerability. Governments can reduce the incidence of children volunteering into armed opposition groups by: (a) not attacking or subjecting to violent harassment, children themselves, their families and homes; (b) where children are involved in armed opposition groups, treating them in accordance with recognised international standards on capture, in particular the Convention on the Rights of the Child, and allowing regular access to all child detainees by the International Committee of the Red Cross. (Chapters 2, 4)

An international standard of 15 years as the minimum age for recruitment into any armed force or armed group and for participation in hostilities is too young (Convention on the Rights of the Child, Article 38, 1977 Additional Protocol I, Article 77(2), 1977 Additional Protocol II, Article 4(3)(c)). Most national law already sets the minimum age for conscription at 18 years (as well as the age of political majority)[420]. International law also recognises 18 as the age of adulthood in many respects. Article 1 of the Convention on the Rights of the Child sets the general age of majority at 18 years. The special status of persons under 18 years is recognised both in relation to the prohibition on the application of the death penalty[421] and in relation to employment or work "likely to jeopardise the health, safety or morals of young persons"[422]. In relation to recruitment, Article 38(3) marks

[420] See Table 2

[421] Convention on the Rights of the Child, Article 37(a), International Covenant on Civil and Political Rights, Article 6(5), Additional Protocol I, Article 77(5), Additional Protocol II, Article 6(4)

[422] International Labour Organisation Convention No. 138 on Minimum Age, 1973, Article 3(1)

out the 15–18 year age group and exhorts states recruiting in that age bracket to give priority to the eldest. The evidence of this report is that children are recruited predominantly because not enough adult recruits are forthcoming, or in order to use them as spies or to commit atrocities. These are not reasons compatible with the Convention on the Rights of the Child. In addition, the evidence of the physical, mental and emotional effects on under-18s of involvement in armed forces or armed opposition groups and participation in hostilities clearly point to the need for the minimum age to be raised. The proposed Optional Protocol to the Convention on the Rights of the Child on involvement of children in armed conflicts should be adopted as soon as possible setting 18 years as the minimum age for any kind of recruitment into any armed forces or armed groups, and for any kind of participation in hostilities. (Chapters 5, 7)

LEGAL AND OTHER ISSUES

Child soldiers are usually treated in the same way as all other soldiers. However, in government armed forces, the treatment of recruits is often inhuman and degrading in the extreme. Such treatment may particularly affect children because they are less robust physically, mentally and emotionally. The general issue of the ill-treatment of recruits and its conformity with international human rights standards needs to be addressed. The UN Sub-Commission on Prevention of Discrimination and Protection of Minorities should undertake a study of the treatment of recruits in government armed forces, with a view to examining its compatibility with human rights standards, as well as the consequences of inhuman and degrading treatment for the recruits and for the subsequent behaviour of the armed forces. (Chapter 4)

Under civilian legal regimes, special provision is normally made at all stages in the legal process to take account of the age of the child. In situations of armed conflict or internal disturbances, often emergency or anti-terrorism legislation is introduced which takes no account of age. Alternatively the age of criminal responsibility is reduced in the face of children's involvement. Similarly, military law, punishment and discipline do not take the impact on children into account. The Committee on the Rights of the Child should consider the questions of applicability to children of (a) emergency and anti-terrorism legislation and (b) military law, punishment and discipline (including in military schools), and should also consider the question of the ages of criminal responsibility in particular in relation to situations of armed conflict, states of emergency and terrorism, and their compatibility with the Convention on the Rights of the Child. Governments should review their legislation to ensure its conformity with the Convention and other international standards. (Introduction, Chapter 4)

In many situations of internal armed conflict, internal strife or disturbances, the government denies that there is an armed conflict, thereby denying the applicability of international humanitarian law to itself and any opposing armed groups, and derogating from its obligations under international human rights law. (a) Where a Government derogates under a human rights treaty, it should not be able to deny the applicability of at least common Article 3 of the Geneva Conventions of 1949. (b) Governments should become parties to Additional Protocol II to the Geneva Conventions of 1949, and governments and opposition groups should be encouraged to apply the Protocol. (Introduction, Chapter 7)

The ready availability of simple-to-operate lightweight automatic weapons has transformed the capacity of children to serve as combatants on something approaching an equal footing with adults. The factors which combine to produce child soldiers – disruption of families, lack of educational and economic opportunities – are exacerbated by the children's participation in armed forces or armed opposition groups. On leaving such service, the lack of other opportunities open to them, together with the acquired skills in weapons, encourages some to seek a future in violent crime. Governments should regulate the production and flow of lightweight automatic weapons. (Chapter 1)

ANNEX I

Issues and questions for the case studies

QUESTIONNAIRE A

1. **Context:** Describe the conflict, including the type of conflict, background, duration, different stages or phases, and how it ended (if it has), in order to set the scene for your answers and to facilitate a general understanding of the situation. An overview of other relevant information would also be helpful, e.g. the percentage of children in the population, the difficulties you have encountered in doing the case study, and the nature of your (or your organisation's) work in the country.

2. **Recruitment:** This section seeks to identify how many and which children become involved in armed forces, how they become involved, what functions they perform initially and whether this changes over time. Particular attention should be paid to the affect of gender in relation to all these issues.

2.1 Numbers in armed forces, with a breakdown of ages (lowest age, under 15, 15–18 years) and sex. How does this relate to the number of adults in the armed forces? At what age are children considered to be adults for this purpose? Is this different for girls? Is age the criterion used or is there another definition, e.g. height? Is there a minimum age for legal recruitment (compulsory or voluntary) under national law? Are those responsible for recruitment aware of the international standards about the minimum age for recruitment?

2.2 Types of activity performed. Does this vary according to age, sex, method of recruitment, type of armed forces, duration of service? Are girls forced/expected/encouraged to render sexual services? Does this vary according to their age and the other functions which they perform?

2.3 Type of armed forces in which children are serving, e.g. governmental, opposition, self-defence forces. Are there armed forces which have no child soldiers? Were children involved in any peacetime armed forces?

2.4 Method of recruitment: compulsory, forced, voluntary, induced? If voluntary, what were the reasons for volunteering? Were benefits offered to encourage volunteers (financial, other)? Are there situations where the armed groups are considered to be better off than the governmental armed forces? Are some categories of children more vulnerable to recruitment than others? Which categories, e.g. poor, less-educated, those from families with problems, children separated from their parents/family (give reasons for separation), certain age group(s), boys/girls, rural/urban, refugees and/or displaced, those in conflict zones, specific ethnic/indigenous or religious groups? To what type of recruitment are they vulnerable and why? Are particular categories targeted for recruitment and if so, which ones, why and by whom? Are certain categories not targeted for recruitment and if so, why?

3. **Treatment in the armed forces:** This section seeks to identify how children are treated or affected while they are in the armed forces. The similarities and differences between children and adults both in treatment and in the effects of their experience should be addressed. Particular attention should be paid to the effect of age and gender in relation to all these issues.

3.1 Are families involved in the armed forces in any way, e.g. cooking for their soldierfamily members, caring for wounded family members?

3.2 Conditions of work: are the children paid and how does this compare with adult soldiers? Does it vary according to age, gender, type or duration of service? Do they get leave, or time off for recreation or visits to their family (do adult soldiers)? Are they given training? Does this include training in any skills which might be useful in civilian life? Are they subjected to toughening-up procedures? What punishments are given for disobedience? Do they receive any education? Do they get preferential feeding or other treatment in comparison with adult soldiers? Are they subject to abusive treatment, such as provision of drugs, alcohol, the rendering of sexual services, other?

3.3 Duration of involvement in armed forces: how long do children stay in the armed forces? What affects the duration of their service? Do the tasks they perform and the treatment they receive vary with the duration of service?

3.4 Numbers killed and injured, and types of injury: how many child soldiers (including those in support functions) get killed? How does this compare with the numbers of adult soldiers killed and with the number of child civilians killed? Do the deaths result from different causes? How many child soldiers are injured? What types of injury do they suffer? How do these compare with the numbers and types of injury to adult soldiers and to child civilians? Are there injuries which are child specific, e.g. hearing loss caused by firing automatic weapons? How do age and gender affect numbers and types of death and injury of child soldiers?

3.5 Provision of treatment for injured: Are injured child soldiers provided with medical care? Are they returned to their families for care? If so, what happens to those without families? Are they abandoned to their own devices? How does their care compare with the provision for injured adult soldiers?

3.6 Numbers and reasons for demobilisation other than at the end of an armed conflict: e.g. invalidity, capture, other? Are there circumstances in which organised demobilisation has taken place other than at the end of the conflict?

3.7 Treatment on capture: How are child soldiers treated when captured? Are they treated as prisoners of war? Is any provision made for access (by families or others)? Is any special provision made for them as children, e.g. education? Are they returned to their families? Are they recruited into the capturing force? Does their treatment differ between different parties to the conflict? Does age, sex, involvement in killings/atrocities, method of recruitment, injuries or other factors affect the way in which they are treated?

QUESTIONNAIRE B

The purpose of this questionnaire is to supplement Questionnaire A by undertaking an in-depth analysis, including observations and suggestions (taking account of gender) of the reasons for the involvement of children as active participants in the conflict and what happens to them afterwards.

4. **Reasons for involvement**: This section is intended to identify the factors which decide why children become involved in some conflicts and not in others, or change from passive observers/victims into active participants. It should include an analysis of methods of recruitment and involvement, including the reasons for volunteering or circumstances of engagement, and information on any preventive strategies employed to reduce or eliminate recruitment of children – whether by those in authority or by communities, families, the children themselves or others – and their effectiveness. Observations, suggestions and recommendations in relation to actions and strategies to decrease or prevent such involvement in the future should be included.

4.1 Have there been different stages in the conflict? How has this affected the involvement of children? At what stage in the conflict did children first become active participants or have they been involved throughout? Historically, have children always been involved in armed conflicts (if so, from what age) or is this something new or different in size or kind?

4.2 How did children become involved? This is an opportunity to add anything to the answers already provided in Questionnaire A.

4.3 What specific factors led to this involvement: e.g. shortage of adult soldiers, the attitude of the military/political leadership, ideology, vulnerability of certain groups of children, the need (or perceived need) to support or protect the family, peer group pressure? Was it a policy decision by those in military/political authority? Do families/communities support the involvement of their children, if so why, e.g. material benefits, ideology/religion/culture?

4.4 What factors inhibited or prevented such involvement, e.g. availability of education, ideology, stable family environment, opposition by families (mothers?) or communities, opposition by military/political/religious or other leaders?

4.5 Were any specific strategies employed to reduce/prevent child recruitment, including voluntarily, if so what, by whom and how effective were they, e.g. keeping schools open, sending children away from the area or out of the country, reuniting separated children with families, making financial or material contributions instead, provision of alternative activities for children (e.g. as first aid workers)?

4.6 How do the military view the involvement of children in their armed forces and their relationship to them? How do they view the involvement of children in opposing armed forces?

5. **The post-service situation**. The purpose of this section is to understand the short-, medium- and long-term effects of children's participation in conflict. In this regard, when answering the following questions it would be most helpful if information describing the situation of former child soldiers could be compared where appropriate with that of former adult soldiers and of children who were not actively involved in the conflict.

5.1 Education: What proportion of former child soldiers are attending school? Is their present level of education appropriate to their age? Are special programmes considered necessary to meet their educational needs? Why, and what form do/will they take? Who will be responsible for their implementation – e.g. government/military; local/international organisations? In addition to enabling the children to regain the education they lost as a result of their involvement in the conflict, are educational activities also considered to assist in a more general way in re-educating the children to civilian life? How is this accomplished? Do teachers identify any problems working with child soldiers? Are teachers given support and assistance in their work? Is it considered that they will require any special training to work with former child soldiers? What is the attitude of the former child soldiers to education?

5.2 Physical and mental well-being: Do former child soldiers have any special health problems – e.g. physical disabilities; sexually transmitted diseases? Are the children considered to behave differently from other children who were not actively involved in the conflict? How do people describe their behaviour? Are these issues specific to certain groups/categories of child soldiers, e.g. girls/adolescents/children from a particular family/ethnic background? Do these health/social issues create difficulties in their daily lives and affect their successful reintegration? What forms of intervention could help in overcoming these difficulties? Is there any indication that former child soldiers are engaged in criminal/violent activities? In situations where some time has passed since their involvement in the conflict, are there indications of long-term social/emotional consequences for the children, e.g. family discord, difficulties in assuming their appropriate roles in society?

5.3 Employment/Vocational Training: Do former child soldiers experience difficulties in finding employment? What factors influence this situation? Where children have come from rural homes, do they usually resume their family's traditional means of livelihood? Are vocational training schemes available for former child soldiers? Are they appropriate to local circumstances – availability of necessary materials, opportunities for employment etc? Are the child soldiers integrated with other children on these training schemes? Are other forms of job training available, e.g. apprenticeship to local craftspeople/assistance to learn the family's traditional means of income generation? Who initiates and is responsible for these schemes? Do personnel on these schemes identify any special problems working with former child soldiers? Is special training/support available to them in their work with the children?

5.4 Community Attitudes and Involvement: Do the children's families and communities identify any social/economic consequences due to their participation in the conflict? Do they affect their reintegration into civil society? How do they cope with these problems? Are some problems specific to certain groups of children – e.g. age/gender-related/family circumstances/knowledge of role played/acts committed by the child during conflict? Is there any community mobilisation to assist the children's reintegration? Who initiated this, parents/teachers/community leaders... ? What measures do the community think could be implemented to facilitate the children's social reintegration? Did the families/community support the recruitment of children and, if so, why? Has their attitude changed towards this form of recruitment and, if so, what brought this about? Is there support for recruitment of children only from certain sectors in the community – who, why, and how do other community members cope with this? Do the families/community identify any benefits to themselves or their children as a result of participation in conflict? What are they?

5.5 The Children: Every effort should be made to include in the case study discussions with child soldiers themselves. These discussions should reflect their attitudes and concerns on the above issues 5.1 to 5.4 in order that we can incorporate their views about the factors that contribute to their reintegration into civil society and the difficulties they are experiencing. Where girls have been involved in the conflict, their special concerns should be noted.

6. **Planning for Reintegration.**

6.1 For countries where conflict has ended. Did the parties to the conflict sign a peace agreement? If so, was the participation of children as soldiers recognised/specified in any way in the peace agreement? Did this recognition reflect concern for the effects of militarisation on children *per se*, and the consequent need to assist them to reintegrate into civil society? If so, what is considered to have influenced the decision to incorporate recognition of this particular group of soldiers in the peace agreement? Have former child soldiers received any demobilisation benefits? What form do they take? Are they comparable to those awarded to adult soldiers? Have programmes for rehabilitation/reintegration of former child soldiers been implemented? What form do they take and who is responsible? What factors are considered (a) to assist this process and (b) to hinder programme development? Are any special groups of former child soldiers targeted as in special need of programmes? Who are they and how are they identified? Are any children excluded from the programme? Is the military supportive of/involved in the children's rehabilitation? Are the children monitored in any way? Who is responsible for the monitoring? What issues are prioritised?

6.2 For countries where conflict is on-going. Have children actively serving as soldiers been officially demobilised or otherwise ceased participating in the conflict? Who are these children – members of government forces/rebel groups/militias/other? Did these children receive any benefits – demobilisation pay/clothing/tools, etc. to assist them on their return to their families/communities? Is their progress monitored in any way? What factors influenced the end of their participation – a specific policy decision/resistance from families and community/pressure from local/international organisations/changes in the nature of the conflict? Is there any planning now for the future rehabilitation/reintegration of children presently involved in the conflict? Who initiated this planning process and which organisations (governmental/local/ international) are involved? What factors influenced the decision to develop such programmes? What factors are hindering the planning process? What are considered to be the priority issues for the children in addressing rehabilitation/reintegration? Are the children involved in the planning in any way?

ANNEX 2

Contributors

Jesuit Refugee Service, Asia-Pacific

International Catholic Child Bureau, Asia

Connell Foley, Cambodia

Colectivo por la Objecion de Conciencia al Servicio Militar, Colombia

Lutheran World Federation, El Salvador

Yitayew Alemayehu, Shoangizaw Chane, Ethiopia

Ismet G. Imset, London, Great Britain

Defensa de la Ninez Internacional, Guatemala

El Comite de Paz y Justicia de la Iglesia Menonita Hondurena y Participaciónes, Honduras

Defence for Children International, Israel

Georges J. Assaf, Lebanon

Children's Assistance Program, Inc, Liberia

Ministerio da Coordenacao de Accao Social, Mozambique

Centre for the Victims of Torture, Nepal

World Vision International, Nicaragua

Mine Clearance Planning Agency, Pakistan

Servicio Paz y Justicia, Paraguay

Rädda Barnen, Peru

Centre for Antiwar Action, Belgrade, Serbia

National Institute for Public Interest Law and Research, South Africa

Images Asia, Thailand

World Vision International, Uganda

Office of the UN High Commissioner for Refugees

ANNEX 3

List of Abbreviations

ABSDF	All-Burma Student's Democratic Front (Burma/Myanmar)
ANC	African National Congress (South Africa)
API	Additional Protocol I of 1977 to the Geneva Conventions of 1949
APII	Additional Protocol II of 1977 to the Geneva Conventions of 1949
BRA	Brugainville Revolutionary Army (Papua New Guinea)
CRC	Convention on the Rights of the Child of 1989
ECOMOG	West African States' Peacekeeping Mission (Liberia)
ELN	Ejercito de Liberación Nacional (National Liberation Army; Colombia)
EPL	Ejército Popular de Liberación (Popular Liberation Army; Colombia)
FAES	Fuerzas Armadas El Salvador (Salvadorean Armed Forces)
FARC	Fuerzas Armadas Revolucionarias Colombianas (Revolutionary Armed Forces of Colombia)
FIS	Front Islamique du Salut (Islamic Salvation Front; Algeria)
FMLN	Farabundo Martí para la Liberación Nacional (Farabundo Martí Liberation Movement (El Salvador)
FPR	Front Patriotique Rwandais (Rwandan Popular Front)
FRELIMO	Frente de Libertaçao de Moçambique (Front for the Liberation of Mozambique)
FRETLIIN	Frente Revolucionaria de Timor Leste (Revolutionary Front of East Timor)
FRUD	Front for the Restoration of Unity and Democracy (Djibouti)
GC	Geneva Conventions I–IV of 1949
GIA	Groupe Islamique Armé (Armed Islamic Group; Algeria)
ICRC	International Committee of the Red Cross
IRA	Irish Republican Army (Northern Ireland)
JSS/SB	Jana Sanghati Samiti/Shanti Bahini (Bangladesh)

KNU	Karen National Union (Burma/Myanmar)
LRA	Lord's Resistance Army (Uganda)
LTTE	Liberation Tigers of the Tamil Eelam (Sri Lanka)
MNLF	Moro National Liberation Front (Philippines)
MRTA	Movimiento Revolucionario Tupac Amaru (Peru)
MTA	Mong Tai Army (Burma/Myanmar)
NGO	Non-governmental organisation
NMSP/MNLA	National Mon State Party/Mon National Liberation Army (Burma/Myanmar)
NPA	New People's Army (Phillipines)
NPFL	National Patriotic Front of Liberia
OAU	Organisation of African Unity
PAC	Pan African Congress (South Africa)
PAC	Patrullas de Autodefensa Civil (Civil Defence Patrols; Guatemala)
PDF	People's Defence Force (Sudan)
PDF	People's Democratic Front (Burma/Myanmar)
PDK	Khmer Rouge (Parti Democratique de Kampuchea; Cambodia)
PLO	Palestine Liberation Organisation
PKK	Kurdistan Worker's Party (Turkey)
RBG	Royal Body Guard (Bhutan)
RENAMO	Resistencia Nacional Mozambique (Mozambique National Resistance)
RUF	Revolutionary United Front (Sierra Leone)
SDUs	(unofficial) Self-Defence Units allied to ANC (South Africa)
SLA	South Lebanon Army
SPLA	Sudan People's Liberation Army
SPUs	(unofficial) Self-Protection Units allied to Inkatha (South Africa)
UDF	Ulster Defence Force (Northern Ireland)
ULIMO	United Liberation Movement (Liberia)
UNHCR	United Nations High Commissioner for Refugees
UNICEF	United Nations Children's Fund
UNITA	Uniao Nacional para a Independencia Total de Angola
URNG	Unidad Revolucionaria Nacional Guatemalteca (Guatemalean National Revolutionary Unit; Guatemala)
UWSP/UWSA	United Wa State Party/United Wa State Army (Burma/Myanmar)

ANNEX 4

Rädda Barnen's Project on Child Soldiers

Rädda Barnen (Swedish Save the Children) has for many years campaigned against the use of child soldiers. In 1995, the organisation launched a new project on child soldiers in line with its previous efforts.

The project has three basic components:

1. A regular newsletter in English.

2. A data base with references to documents on child soldiers.

3. An international network of concerned organisations and individuals.

Children of War, the newsletter, is published quarterly and has so far been issued four times. Its main purpose is to raise public opinion and stimulate action against the use of child soldiers. Each issue has a campaign section, which up to now has focused on the work within the UN to draft an optional protocol to the Convention on the Rights of the Child, raising the minimum age for military recruitment and participation. Children of War also contains media abstracts, with recent media reports on children bearing arms, and information on the work by various organisations in relation to child soldiers.

The newsletter mailing list contains over 1,600 addresses, of which more than half are located outside Europe and North America. E-mail subscription is possible, and all issues of the newsletter are available at the

World Wide Web on Rädda Barnen's home page (see addresses below).

The second main component of the project is the data base on child soldiers, *SOLDOC*. It contains references to documents with information on child soldiers, the militarization of children and peacetime military use of children. Rädda Barnen's ambition is to cover books/reports that deal exclusively with child soldiers as well as books/reports printed in 1990 and later that to a large extent deal with child soldiers. The data base will also contain references to news articles, UN documents, videos and other documentation of interest. All documents in the data base are read and supplied with key words, in order to make them searchable. In 1997, the data base will be made available on-line via the Internet. For now, requests for searches must be addressed directly to the project coordinator. Also in 1997, SOLDOC will be complemented with a second data base containing numbers, ages and other factual information on child soldiers, organised by country.

As a natural and necessary component of the overall project, *an international network* of interested organisations and individuals is evolving. As mentioned above, the register of addresses for Children of War contains over 1,600 addresses. Of them, around 100 are categorised as "child soldier network" contacts, that is, individuals with whom Rädda Barnen has had a direct contact concerning child soldiers. One of the long-term objectives of the project is to increase the awareness of the child soldier issue among humanitarian organisations with field offices.

It should be underlined that Rädda Barnen's Project on Child Soldiers is not intended to compete with similar work by other organisations. Accordingly, Rädda Barnen is looking for the widest possible co-operation.

For more information, please contact:

AnnaLena Andrews
Project Co-ordinator/Editor
Rädda Barnen's Project on Child Soldiers
Rädda Barnen, 107 88 Stockholm, Sweden
Tel: +46 8 698 90 92, Fax: +46 8 698 90 13
E-mail: al.andrews@baf.rb.se
WWW-address: http.//www.rb.se/

Bibliography

For further information on the literature listed below, including the 26 case studies produced for the book, please contact Rädda Barnen's Project on Child Soldiers (see address in Annex 4). The case studies are listed in Table 1 in the Introduction.

Amnesty International, *Sierra Leone: Human rights abuses in a war against civilians.* London, Sept. 13, 1995.

Amnesty International, *Sierra Leone: Prisoners of war?: Children detained in barracks and prison.* London, Aug. 12, 1993.

"Angola: Civilians devastated by fifteen year war." In: *News From Africa Watch.* 1991 (vol. 3), no. 1.

Balian, H., *Armed conflict in Chechnya, its impact on children.* Covcas Center for Law and Conflict Resolution, Arlington, Va., 1995. (Report prepared for the UN study on the Impact of Armed Conflict on Children).

Blomqvist, U., "Social work in refugee emergencies: capacity building and social mobilisation: the Rwanda experience". In: McCallin, M. (ed.), *The psychological well-being of refugee children: research, practice and policy issues* (second edition), International Catholic Child Bureau, Geneva, 1996.

Bonnerjea, L., *Family tracing: a good practice guide.* Save the Children, London, 1994. (Save the Children Development Manual no. 3).

Boothby, N., Upton, P. & Sultan, A. "Boy soldiers of Mozambique." In: *Refugee Participation Network.* 1992: no. 12 (March).

"Boy soldiers: a new and ruthless breed of warriors." In: *Newsweek,* 1995 (vol. CXXVI), no. 6.

Children, genocide and justice: Rwandan perspective on culpability and punishment for children convicted of crimes associated with genocide. Save the Children Federation USA, Kigali, 1996.

Children in war: community strategies for healing. Save the Children USA, University of Zimbabwe and Duke University, Durham, N.C., 1995.

Defensoría del Pueblo (Human Rights Ombudsman), *Victimas de la Violencia: El conflicto Armado en Colombia y los Menores de Edad* (Victims of Violence: The Armed Conflict in Colombia and the Minors), Bogota, Colombia, May 1996.

Dütli, M. T., "Captured child combatants." In: *International Review of the Red Cross,* 1990: no. 278 (September-October).

Fozzard, S., *Surviving violence: a recovery programme for children and families.* International Catholic Child Bureau, Geneva, 1995.

Gibbs, G., "Post-war social reconstruction in Mozambique: reframing children's experience of trauma and healing". In: *Disasters,* 1994 (vol. 18), no. 3.

Goodwin-Gill, G. & Cohn, I., *Child soldiers: the role of children in armed conflicts.* Clarendon Press, Oxford, 1994.

Hampson, F.J., *Legal protection afforded to children under international humanitarian law.* University of Essex, Colchester, 1996. (Report prepared for the UN study on the Impact of armed conflict on children).

Human Rights Watch/Africa, Arms Project, *Angola: Arms trade and violations of the laws of war since the 1992 elections.* New York, 1994.

Human Rights Watch/Africa, Children's Rights Project, *Children of Sudan: slaves, street children and child soldiers.* New York, 1995.

Human Rights Watch/Africa, Children's Rights Project, *Easy prey: child soldiers in Liberia.* New York, 1994.

Jareg, E., *Rehabilitation of child soldiers in Mozambique.* Redd Barna, Oslo, 1993. (Note prepared for the Secretary General of Redd Barna).

Jareg, E. & Jareg, P., *Reaching children through dialogue.* Macmillan Press, Basingstoke, 1994.

Jareg, E. & McCallin, M., *The rehabilitation of former child soldiers: report on a training workshop for caregivers of demobilised soldiers: Freetown, Sierra Leone, September 1st to 3rd 1993.* International Catholic Child Bureau, Geneva, 1993.

Jongman, A.J. & Schmid, A.P., "World conflict map, 1994–1995." In: *PIOOM Newsletter and Progress Report.* 1995 (vol. 7), no. 1.

Maramba, P., *Reintegration of demobilized combatants into social and economic life: the Zimbabwean experience.* Geneva, 1995. (Report on a consultancy for the ILO).

McCallin, M., *The reintegration of young ex-combatants into civilian life: expert meeting on the design of guidelines for training and employment of ex-combatants, Harare,*

11–14 July 1995. International Labour Office, Geneva, 1995.

Price-Cohen, C., "Considerations affecting the implementation of the United Nations Convention on the Rights of the Child in situations of forced migration." In: M. McCallin (ed.), *The psychological well-being of refugee children: research, practice and policy issues* (second edition). International Catholic Child Bureau, Geneva, 1996.

Reichenberg, D. & Friedman, S., "Traumatized children: healing the invisible wounds of children in war: a rights approach." In: Danieli, Y., Rodley, N. S. & Weisaeth, L. (eds.), *International responses to traumatic stress*. Baywood Publishing Company, Inc., Amityville, N.Y., 1996.

Ressler, E., Tortorici, J. M. & Marcelino, A., *Children in war: a guide to the provision of services*. UNICEF, New York, 1993.

Roberts, S. & Williams, J., *After the guns fall silent*. Vietnam Veterans of America Foundation, Washington D.C., 1995.

Save the Children Alliance, *Promoting psychosocial well-being among children affected by armed conflict and displacement: principles and approaches*. Geneva, 1996. (Paper prepared as a contribution to the UN study on the Impact of armed conflict on children).

Tolfree, D., *Restoring playfulness: different approaches to assisting children who are psychologically affected by war or displacement*. Rädda Barnen, Stockholm, 1996.

Tortorici, J.M., "Peace education in Nicaragua." In: M. McCallin (ed.), *The psychological well-being of refugee children: research, practice and policy issues* (second edition). International Catholic Child Bureau, Geneva, 1996.

United Nations, *Comments on the report of the Working Group on a draft optional protocol to the Convention on the Rights of the Child on involvement of children in armed conflicts*. Geneva, 1995. (UN document E/CN.4/1996/WG.13/2, 23 Nov.)

Further reading drawn from the Rädda Barnen database

For further information on the literature listed below, please contact Rädda Barnen's Project on Child Soldiers (see address in Annex 4).

Aldrich, G.H. & van Baarda, Th.A. (eds.), *Conference on the rights of children in armed conflict: final report of a conference held in Amsterdam, the Netherlands, on 20–21 June 1994*. International Dialogues Foundation, The Hague, 1994.

Audoin-Rouzeau, S., *La guerre des enfants 1914–1918: essai d'histoire culturelle*. Armand Colin, Paris, 1993.

Boothby, N., *Reclaiming the killing field.* Duke University, Durham, N.C., 1990. (Dukes Perspective).

Boothby, N., Upton, P. & Sultan, A., *Children of Mozambique: the cost of survival.* U.S. Committee for Refugees, Washington, D.C., 1991, (Issue paper).

Center on War and the Child, *Afghanistan: The war against children: a summary report.* Eureka Springs, Ark., 1987.

Center on War and the Child, *The children of Mozambique's killing fields: a summary report.* Eureka Springs, Ark., 1989.

Center on War and the Child, *The recruitment and use of children in the Gulf war: a summary report.* Eureka Springs, Ark., 1988.

Center on War and the Child, *Uganda: land of the child soldier: a summary report.* Eureka Springs, Ark., 1987.

"Children in combat." In: *Human Rights Watch Children's Project.* 1996 (vol. 8), no. 1(G).

Children of war: a newsletter on child soldiers from Rädda Barnen. 1995, no. 1- . Rädda Barnen, Stockholm, 1995- .

David, P., *Enfant sans enfance.* Hachette, Paris, 1995.

Dodge, C. P. & Raundalen, M., *Reaching children in war: Sudan, Uganda and Mozambique.* Sigma Forlag, Bergen, 1991.

Haapiseva, J., *Report of a field trip to Liberia, Mozambique and Zimbabwe on behalf of the Lutheran World Federation.* Lutheran World Federation, Geneva, 1994.

Hedlund Thulin, K. (ed.), *Children in armed conflict: background document to the plan of action concerning children in armed conflict.* Henry Dunant Institute, Geneva, 1995.

Jareg, E., *From soldiers to citizens: reintegration of children involved in war.* Redd Barna, Oslo, 1994.

Kent, G., *Implementing the rights of children in armed conflicts.* University of Hawaii, Spark M. Matsunaga Institute for Peace, Honolulu, 1992. (Working Paper).

Kosonen, A., *The special protection of children and child soldiers: a principle and its application.* University of Helsinki, The Institute of Public Law, Helsinki, 1987.

Lindsay-Poland, J., *Youth under fire: military conscription in El Salvador: a summary report.* Center on War and the Child, Eureka Springs, Ark., 1989.

Louyot, A., *Gosses de guerre.* Robert Laffont, Paris, 1989.

McCallin, M., *Community involvement in the social reintegration of child soldiers.* London, May 1996. (Paper presented at the Save the Children (UK) conference on rethinking the trauma of war).

McCallin, M. & Jareg, E., "The rehabilitation and social reintegration of child soldiers. "In: McCallin, M. (ed.), *The psychological well-being of refugee children: research, practice and policy issues (second edition).* International Catholic Child Bureau, Geneva, 1996.

Parker, R., *Iran's child martyrs: a summary report.* Center on War and the Child, Eureka Springs, Ark., 1987.

Pint, N., *Child Soldiers.* University of Essex, Colchester, 1993. (Dissertation).

Racsmány, Z., *Sacrified by culture: the role of cultural factors in understanding the phenomenon of child soldiers.* Uppsala University, Department for Peace and Conflict Research, Uppsala, 1995. (Academic paper).

Raundalen, M. & Dyregrove, A., *Reaching children through the teachers: a manual for helpers of war-affected children.* University of Bergen, Bergen, 1992.

Raoul Wallenberg Institute of Human Rights and Humanitarian Law, *Children of war: report from the conference on children and war, Stockholm 31 May – 2 June 1991.* Lund, 1991.

Rädda Barnen, *Child soldiers: a report from a seminar by Rädda Barnen and the Swedish Red Cross 12–13 February 1994, Stockholm, Sweden.* Stockholm, 1994.

Rädda Barnen, *No child soldiers!: how article 38 in the Draft Convention on the Rights of the Child could be amended to provide real protection for children in armed conflicts before its adoption by the UN General Assembly.* Stockholm, 1989.

"The lost boys: child soldiers in Southern Sudan." In: *Human Rights Watch/Africa.* 1994 (vol. 6), no. 11.

The state of the world's children 1996. Oxford University Press for UNICEF, New York, 1995.

The University Teachers for Human Rights (UTHR), *Children in the North-East War: 1985–1995.* Jaffna, 1995, (Briefing no. 2).

UNHCR, *Refugee children: guidelines on protection and care.* Geneva, 1994.

UN Economic and Social Council (ECOSOC), *Rights of the child: report of the working group on a draft optional protocol to the Convention on the Rights of the Child on involvement of children in armed conflicts on its second session.* Geneva, 1996. (UN-document E/CN.4/1996/102, Mar. 21)

Woods, D., *Children bearing arms: a form of child labor?: a summary report.* Center on War and the Child, Eureka Springs, Ark., 1990.

Woods, D., *Children bearing military arms in Latin America: a summary report.* Center on War and the Child, Eureka Springs, Ark., 1990.

Woods, D., *Children bearing military arms in Uganda: a summary report.* Center on War and the Child, Eureka Springs, Ark., 1990.

Woods, D. (ed.), *Child soldiers: the recruitment of children into armed forces and their participation in hostilities.* Quaker Peace and Service, London, 1993.

Woods, D., *Lebanon: children in armed conflict from 1975–1989: a summary report.* Center on War and the Child, Eureka Springs, Ark., 1990.

Index

For countries, see also Table 2 (page 50)